A Return to Common Sense

John Ikerd

Edwards

ISBN: 978-1-930217-17-1

○
Edwards   www.rtedwards.com

For order information or reprint permission, contact:
R.T. Edwards, Inc., P.O. Box 27388, Philadelphia, PA 19118 USA.

Cover Photo: A. Huszti/bigstockphoto.com

Library of Congress Cataloging-in-Publication Data

Ikerd, John E.
A return to common sense / John Ikerd.
    p. cm.
Includes bibliographical references and index.
ISBN 978-1-930217-17-1 (alk. paper)
1. Social change. 2. Social justice. 3. Economic policy. 4. Environmental justice.
5. Sustainable development. I. Title.
HM831.I38 2007
303.48'401--dc22
                            2007005387

    1 2 3 4 5 LS 7 8 9 10 11
Printed in the United States of America

Contents

Preface

I believe humanity is in the midst of a great transformation, moving out of the industrial era and into a new and fundamentally different era of human progress. This new post-industrial stage is yet to be given a permanent name. It is being ushered in by the sustainability movement, which includes sustainable development, sustainable living, sustainable agriculture, sustainable fisheries, sustainable forestry, and sustainable almost any other aspect of our economy or society. The basic principles of the sustainability movement are shared by advocates of environmentalism, energy conservation, and the prevention of global warming; of social justice, civil rights, and human rights; of non-violence, nuclear disarmament, and world peace; and of holistic medicine, holistic education, and spirituality.

The size of this new movement is generally underestimated, because the people in any one of these advocacy groups simply do not realize their basic principles are so widely shared with those in related movements. Those who need documentation of the emergence of this new culture in America might want to read *The Cultural Creatives, How 50 Million People Are Changing the World* by Paul Ray and Sherry Anderson.[1] I am convinced the sustainability movement now includes at least one-third of all North Americans and even greater proportions of Europeans and Asians, based on a variety of published sources and personal observations. It perhaps is as strong or stronger on other continents as well, but I simply don't have enough information to be sure. I am certain, however, that the movement continues to grow.

The great transformation is being driven by the growing realization that the neo-classical economic development paradigm, which currently dominates all industrial economies, quite simply is not sustainable over the long run. Paradigms are the mental models or frameworks we use to organize knowledge and information, develop plans and strategies, and to make and evaluate decisions. The industrial economic paradigm is very productive in the short run, but its productivity is supported by the extraction and exploitation of both the natural and human resources upon which its long run productivity inevitably depends. It is using up nature and it is using up society, and when these natural and human resources are gone, there will be no means of sustaining the economy. A growing number of people understand that to sustain civilized human life on earth, we must develop and implement a fundamentally new and different paradigm of economic development.

This great transformation will be accomplished through a new revolution – a period of rapid, and perhaps violent, societal change. Today we are in the midst of a revolution of new ideals – a battle for the hearts and minds of people. If we lose this peaceful revolution of ideals, humanity eventually will be confronted with a violent revolution among desperate people, when the earth's natural resources are no longer adequate to support its numbers and the earth's social resources are no longer adequate to find peaceful ways to allocate what little is left.

The new post-industrial era will require fundamentally different ways of thinking, choosing, organizing, and governing. It will require a different worldview – seeing the world as a complex, living ecosystem, rather than as a big, complex machine. It will require new and different paradigms for science, economics, and society. During such great transformations, the old paradigms are no longer relevant but the new ones are not yet fully developed. Our beliefs of what is real or imagined, true or false, right or wrong, and good or bad are a reflection of our worldviews and our mental paradigms. During great transformations, our means of reaching consensus is by relying on our common sense – our shared insights into what we know in our hearts to be real, true, right, and good. Our differing cultural and religious *values* tend to separate us but our common understanding of fundamental rightness and goodness hold us together. Our intelligent insight, our common sense, must provide the foundation of first principles upon which we can create a new science, a new economy, and a new society. During these times of economic, social, and ecological revolution, we have no choice but to return to our common sense.

Revolutions *begin* with recognition and understanding of the enemy, the villain, which today is the neoclassical economic paradigm of industrial development. However, revolutions are *won* by recognizing and understanding the new and better world to come, the vision, which is the new sustainable paradigm of economic, social, and ecological development. A growing number of people are beginning to understand that the old world is not sustainable. This is a necessary first step. Revolutions are sparked by the necessity for change, but revolutions are won by a commitment to a vision of something fundamentally better. I expect the name of the new era to reflect the new vision of the better world of the future, rather than the lack of sustainability in the world of today.

The new vision is rooted in a more enlightened concept of self-interest. The old self-interest is individualistic, materialistic, short run, and narrow. It's about how to get as much *stuff* as we can for ourselves right now, and trusting the invisible hand of free markets to transform our greed into good. The new self-interest is interpersonal, multidimensional, long run, and inclusive. It's about creating and maintaining relationships that will meet our physical, social, and spiritual needs; it's about making conscious, purposeful decisions to care about others and to care for the earth, as well as for ourselves. The new self-interest recognizes that it is not a sacrifice to care about other people or to care for the earth because these things add purpose, meaning, and happiness to our lives, right now. I have suggested it might be called the "age of insight," as we will have moved beyond logic and reason, by utilizing our uniquely human capacity for intelligent insight. The name doesn't matter, the new vision does.

Finally, revolutions are won by people of hope. Hope doesn't mean that we believe the victory will be easy or quick, or even that the odds are in our favor. Hope instead means that we know what we are pursuing is fundamentally right and good and that victory is possible. Hope is contagious; it spreads from one hopeful person to another, until it explodes into an epidemic of hope, which provides the revolution with the energy it ultimately needs for victory. I don't know whether or not this revolution will be won, but I know that the vision we are pursuing is not only necessary for the future of humanity, but perhaps more important, it is right and good. It makes common sense.

We were put on this earth not only to take care of ourselves, but also to take care of the earth, and to care about each other. Carrying out our purpose in life is the only thing that gives true quality to our life – that makes us truly happy. We will not find true happiness and quality in our lives until we find the courage to return to our common sense and to win the new social, economic, and spiritual revolution.

Endnote

[1] Paul Ray and Sherry Anderson, *The Cultural Creatives: How 50 Million People are Changing the World* (New York: Three Rivers Press, 2000).

Acknowledgements

Much of this book was written in the fall, winter, and spring of 2000-2001. I had just retired, after 30 years in various professorial positions at four different state universities. I had reason to be concerned about my physical health and wanted to put some of the things I had learned over the years into words before I died. The book I wrote at that time was about almost everything – philosophy, science, economics, society, ecology, self-help, culture, spirituality, plus a few other subjects. I liked the book when it was finished and a couple of publishers showed some interest. But the book didn't fit into any particular market category, and it was apparently too personal for many readers.

After a dozen or more rejections, I put the book on my university website, where it was freely available. Each chapter was being accessed an average of about 150 times per month. In the summer of 2006, I received an email message from James Edwards, of R.T. Edwards Inc., the publisher of this book. In a subsequent telephone conversation he told me that he had read the book, was appreciative of the systems thinking evidenced in my writing, and was interested in publishing it. I agreed to the proposition and decided to edit the book down to about 200 pages, to make it more affordable and marketable. We also agreed that we were both more interested in promoting new ideas among the greatest possible number of readers, rather than making a lot of money, but hoped that we might at least cover our out of pocket cost of this publishing venture. I hope we succeed in the former, even if not in the latter. I took out any personal stories that were not essential to my message and made a few revisions to clarify specific points.

So, I first wish to express my appreciation to James Edwards and all of the R.T. Edwards, Inc. crew for having enough confidence in my ideas to help me share them with a new and hopefully broader audience. Since this book is still a story of my life as well as my thoughts, I also want to thank all of the important people in my life, including my friends, scattered all across the continent, and my immediate and extended family, most of which are now back in Missouri. In particular, I want to thank my wife, Ellen, who proofs virtually everything I write, including several versions of this book, provides insightful and valued advice, and gives me the moral support that is absolutely essential to my continued efforts to do my part to help create a better world. I am also thankful that I have been fortunate enough, for whatever reason and in spite of 30 years in academia, to *return to common sense.*

Introduction

1

AN AWAKENING

Awaking from a restless dream, my eyes opened slowly. My mind had been proofing a paper, or something of that nature, when I began to return to consciousness. I looked around. I wasn't in my bed or anywhere else that I might have taken a nap. I was lying on an examining table in a hospital. Someone was asking me how I was doing – if I was okay. Slowly, I began to remember where I had been before I "fell asleep." Someone had been asking me how I was doing then – I had said fine, until the last time he asked. The last time, I had felt a bit light-headed. I remember looking at the heart monitor and seeing the lines tracing the rhythms of my heart going crazy. That's the last thing I remember from my old life. I had awakened to a new life and a new world.

The doctor now asked if I knew what had happened. I didn't. A young nurse who had been assisting with my stress test looked a bit stressed herself – pretty shaken in fact. She had come in to relieve the regular nurse who was called away for mock "code" training – to practice bringing people back to life after their hearts stop beating. My nurse had just practiced the same thing by doing the real thing, for her first time. She would come by my hospital room several times later that week to take electrocardiograms. She always had a strange look in her eyes – as if my being alive was somehow unnatural, or at least unexpected.

The doctor eventually told me what had happened. I had successfully completed my stint on the treadmill – twelve minutes plus, as I recall. It was a routine six-month checkup after what had appeared to be successful angioplasty the summer before – the summer of 1997. I had been sitting on the table, cooling down, when suddenly my heart went into ventricular fibrillation – a spasm of the lower chambers. There is only one hope for survival of ventricular fibrillation – the paddles. Since the crash cart was only ten feet away, I was "dead" for only a minute or two before they shocked me back to life. Had it happened anywhere else, say during my daily jog, I would have stayed dead, forever.

1

Even after I found out what had happened, I felt strangely calm. Why was I in a hospital when my heart stopped? Maybe it just wasn't my time to die, as my mother would have said. I must have something left to do here, I thought. I eventually concluded that my near-death experience was a message: "You're not going to die yet, but you're certainly not going to live forever." If I expected to finish what I was supposed to do here, I now understood I had better get on with it.

I went through open-heart surgery, three bypasses, with no sense of fear. It made no sense that I had come back to life only to die a few days later. I actually enjoyed the days and weeks of my recovery, even with the aches and pains that go with having your chest cracked open and your "innards" rearranged and repaired. My older brother, Tom, came to stay with me for a week. We had planned to take a trip to North Carolina that week to play golf and hike in the mountains; instead, he helped me recuperate. We talked and took lots of walks, and I often napped – as prescribed by my doctors.

After Tom went home, I spent time reading and reflecting on the purpose of my new life. What was I supposed to do that was so important that I had to die to become motivated to get on with it? I had always believed that each life has a purpose, but even at 58 years old, I still had not figured out the purpose of mine.

By chance, I had just checked a book out of the University library – *The Life and Major Works of Thomas Paine*.[1] Paine was a writer during the American and French Revolutions, and a friend of Benjamin Franklin and Thomas Jefferson. He wrote about philosophy and politics in a style that was understandable to the "common man" of his time. At Valley Forge, George Washington ordered that a Paine essay be read to the troops, reviving the spirit of American Revolution during its darkest hour.

Paine's most famous writing was a fifty-page pamphlet called *Common Sense*.[2] In it he wrote that it made no sense for the American Colonies to remain attached to Great Britain, so they should declare their independence and form their own Republic. Paine's words gave voice to yearnings for freedom and gave hope when the cause of revolution seemed hopeless. Throughout the revolution, Paine continued to use the pen name, "Common Sense."

Paine's writing may have been easy reading by standards of the 1770s, but it wasn't easy reading for me. Under any other circumstances, I probably would never have finished the book. But I felt a link somehow between the book and my near-death experience; my having checked this book out at this particular time was more than coincidence. Tom Paine's purpose in the 1770s was somehow related to mine today.

My colleagues at the University would have scoffed at my search for purpose or meaning. In science, all things happen for purely logical and rational reasons. For every effect, there is a logical, rational cause. My heart stopped beating because calcium deposits had blocked much of the blood flow through three of its arteries. My lifestyle reflected virtually none of the high risk factors for heart disease, but there were other factors to consider, such as heredity and internal stress. My heart had stopped for purely physical reasons. The fact that I had forgotten about pretest restrictions on caffeine and had drunk four cups of regular coffee and a tall Starbucks that morning was carelessness rather than destiny. I also had survived by pure chance. If I had gone for a jog rather than a stress test that morning, I would still be dead. I was alive only because I was lucky. I had checked out Thomas Paine's book because I wanted to read his paper on agrarian reform; that was the only reason I had his book at the time. There was no specific purpose or meaning in anything that happened that day. In science, things happen for purely logical, rational reasons; that's all there is to it.

Although a scientist, I had abandoned this "rational explanation for all things" philosophy some time back. In fact, I had been drifting farther and farther from the logical and rational thinking of my colleagues for at least a decade. The near-death experience was only the latest, albeit the most dramatic, of a series of events that had been guiding my life in a different direction for some time. For example, I had slowly changed from being a conservative, free market, bottom line, sort of agricultural economist to something fundamentally different – just how different I didn't yet know.

One of the earliest legs of my journey away from scientific rationality had been my questioning of the difference between knowing *how* and knowing *why*. I knew that all of the logical, rational explanations for why my heart stopped, and why it started again. But, I knew also that blocked arteries and muscle spasms actually didn't explain *why* it stopped, but rather *how*. As with all purely logical explanations of *how* things happen, rationality left me without a clue as to *why*.

In my early adult years, I saw science as the means by which humankind was slowly removing the mysteries of life. Science was born during a time when people thought God was the answer to nearly every question. Why did the earth exist? Because God created the earth. Why did it rain? Because God made it rain. Why were people born? Because God gave them life. In the 1600s, all wisdom was thought to reside in the minds of the religious scholars or clerics. Early scientists, such as Galileo, were condemned by the church because their scientific findings seemed to conflict with religious beliefs of the time. Eventually, however, science

prevailed. As science answered more and more questions, God became
the answer to fewer and fewer.

Scientists claim that the earth was created by a "big bang" in the universe, not by God. It rains because moisture in the air cools as it rises,
condensing into clouds, and eventually into raindrops that fall back to
the earth. A fetus is formed when a sperm fertilizes an egg in the womb,
and about nine months later a child is born. These things happen
because of physical cause and effect relationships, not because of God.
As science discovers more and more of these relationships, fewer and
fewer mysteries were left in the realm of God. So I reasoned that science
eventually would completely eliminate the realm of God. In essence, science was the means by which "man would become his own God."

Confronted with this logical conclusion to this line of thought, however, I began to question my thinking. Science might describe *how* the
earth was formed, but *why* was it formed? What is its purpose or reason
for its being? Science may describe how raindrops are formed, but why
does it rain - what is the purpose or reason for rain? Scientists may
answer: rain provides water for people, it feeds crops and crops feed
people, but what is the reason or purpose of people? Why are people
born? The reproduction process only describes how, not why. Why do
people die? The fact that our heart and brain start and stop functioning
only describes how, not why we live or die.

Eventually, I came to realize that the whole of science, for over four
hundred years, had not given us one single clue to the ultimate purpose
or meaning of anything - certainly not human life. Only far later would
I realize that modern science actually denies the existence of anything
such as purpose. Science asserts that all events are due to the interaction
of matter and motion, acting as if by blind necessity, according to invariable sequences of causes and effects, which scientists identify as laws.[3]
Thus, when it comes to answering the questions of *why*, we are no nearer to the answers today than we were 400 years ago. Science has allowed
humanity to do some wonderful things. But, we will have to look elsewhere for the answers as to *why* things happen. And I would have to
look beyond logic and reason to discover *why* I was still alive.

One thing "dying" most certainly will do is make you stop and think.
In fact, that may be the single most important reason for major crises in
life, although no one will ever be able to prove it. People seem so busy
living, rushing from one minute of their life to the next, they never take
time to stop and think; "why am I doing this?" Even if we ask, we never
get beyond the logical and rational reasons that deal only with *how*. "I'm
doing it for the money," we say. Or maybe we get beyond the money and
say, "I'm doing it for my children." But, why are you doing *this* or *that* for

your children? Why do we choose making money, rather than spending time with our children, as the means of doing something for them? What are we really trying to accomplish with our lives?

In fact, we live mostly by habit. We started doing some things because we were taught to do them, then we tested them for ourselves, and at least some of what we learned seemed to work for us. So, we kept doing them until they became habits. We adjust our habits now and then along the way, based mostly on what seems to be working and what doesn't. But, we rarely look beyond our day-to-day experiences for guidance. In a sense, we all have become scientists; we form hypotheses by observing things around us and then draw conclusions by observing the result of our experiments. But our science of day-to-day living, like science as a profession, can answer only questions of *how*, not *why*.

If we just think about it, it's clear that the purpose of anything, including our lives, can be discerned only by seeing it within the context of the larger whole of which it is a part. Everything is a part of something larger and everything is made up of parts, which are smaller. And, the purpose of any part of a thing is inherently dependent upon the purpose of the larger whole.

For example, the purpose for a line, a number, or a color on a piece of paper becomes apparent only when we look at the other lines, numbers, or colors on the same piece of paper. A line may be the horizon in a painting of a landscape, a number may represent millions of dollars in a financial report, and red may be the color of a movie star's lips in a magazine photo. But, we can determine none of these things without seeing the parts within the context of the whole. Likewise, a doctor may be able to describe the function of a human heart or brain – to transport blood or to process electrical impulses – but the purpose of these organs cannot be determined without considering the body as a whole. The human body is the context – the whole within which the heart and brain derive their purpose or reason for functioning.

The parts of a thing cannot acquire meaning from the other parts of the same thing because the parts have no unique meaning apart from the whole. The nature of the whole of anything depends on the arrangement or organization of its parts, not just on the nature of the parts that are arranged or organized. The essence of the whole of a thing is not just in its parts, but also in the arrangement of its parts. Thus, each part only has meaning within the context of a particular whole – as part of a particular arrangement.

For example, the parts of the human body, laid out side by side on a surgeon's table, have no particular meaning or purpose, either individually or in total. They take on meaning only when arranged in that unique and mysterious way that allows them to take on human life. In this case, the organization is sequential as well as spatial. If the organs fail to function together, in a specific, prescribed sequential order, the body dies. The whole is fundamentally changed. The same parts arranged in any other way, either spatially or sequentially, would not form a living human being, and thus, would have a very different meaning. The parts of the human body gain their meaning not from each other, but instead from the whole of their unique arrangement.

Likewise, a person cannot possibly derive the meaning or purpose of their life from the simple process of living. Experimenting and observing can only provide us with information concerning how we relate to other parts of our universe, to the other people and other things with which we interact. Our growing understanding of causes and effects leaves our life without understanding of purpose or meaning. The purpose and meaning of our life cannot be derived from our relationships with other people or other things around us because we are part of the same whole.

We may know that the colors red, blue, and yellow can be used to create all other colors, and with these three colors, we can paint any picture we might choose. But our knowledge of colors can't tell us which picture we should paint or the purpose of our painting it. We may learn which functions we are most capable of performing – we may be good planners, workers, teachers, or painters. But knowing our capabilities can't tell us the purpose of our work or our reason for living. The meaning of our individual lives must be derived in turn from the purpose of the whole of human life, and we simply are not physically capable of observing that greater whole of which we humans are but a part.

For us humans, the essence of this greater whole will forever remain intangible and abstract. Plato argued that one can never gain "pure knowledge" through observation because anything that can be observed is always changing, whereas, pure knowledge never changes.[4] He argued that we observe only imperfect examples of the true "form" of things – the order or architecture of pure knowledge. We can never fully grasp pure knowledge through observation, because it exists only in the abstract. On the other hand, Plato argued that when reason is *properly used* the results are intellectual insights that are certain, universal, and eternal – the forms or substances that make up the real world.

Using Plato's terminology, the purpose of our life is derived from the universal, eternal "forms" or substances of things – from the order or architecture of pure knowledge. Our intellectual insight tells us there must be some higher level of organization or order of things; otherwise, our lives could have no purpose or meaning. Without purpose, it would make no difference what we did to ourselves, to others, or to the world around us. Nothing would be fundamentally right or wrong, so any action would be equally good or bad. Since this conclusion is clearly inconsistent with our intuitive understanding of reality, we conclude there must be some higher order of things – some larger, universal whole of which we are but a part.

We can never prove this conclusion scientifically, however, because this higher order exists at a level beyond the realm of our direct observation or experimentation. We can see evidence of this order reflected in the world around us and in the lives of other people. However, our *changing* observations have meaning only because we have some intuitive *knowledge* of the *unchanging* "form" and unseen order of things. We observe only imperfect examples. Thus, *knowledge* of the purpose and meaning of life must necessarily come from insight, intuition, or more clearly, from our *common sense*.

Our common sense is our intellectual insight into the nature of pure knowledge. Through our common sense, we can visualize in our minds the higher level of organization. Through observing the interrelationships among things in nature, including human relationships, we can gain insights into the nature of the higher order. But we can never fully validate our insights through science, because the order exists only in the abstract.

We can't find purpose through experimentation and observation, but we can find it through insight and intuition – by searching with our soul as well as with our mind. Our common sense provides the intellectual foundation for our understanding of right and wrong as well as truth and reality. We must access our common sense through our spiritual sense of knowing, not through intellectual inquiry. In searching for pure knowledge, we must rely on that part of us that transcends our physical selves – our metaphysical or spiritual selves. The spiritual and metaphysical are not the exclusive realms of those who meditate or devote their lives to deep thought. A common spirit lives in us all.

We all have a shared, or common, sense that our life has purpose, because we all derive our purpose from the same order of things. We are all parts of the same whole. Our common sense of purpose comes to us all quite naturally, regardless of our native intelligence, our social status, our background, or our training and education. This part of our sense of knowing can be called *common* because people in general share this *same* sense.

Common sense has become an overused and often abused colloquialism, but the concept has deep philosophical roots. Thomas Reid, nineteenth-century Scottish philosopher wrote, "All knowledge and science

must be built upon principles that are self-evident; and of such principles every man who has common sense is competent to judge."[5] These
self-evident principles provide a starting point, and lacking a starting
point, all logic and reasoning eventually become circular and thus use-
less. For example, first principles of algebra, called axioms or laws, are
the foundation for all mathematical proofs. One such axiom is, *a* times *b*
equals *b* times *a*. This may seem obvious, but such is the nature of all
common sense.

The philosophy of common sense does not reject science as a
means of gaining understanding. Thomas Huxley, a noted English
botanist, once wrote, "All truth, in the long run, is only common sense
clarified."[6] When Albert Einstein wrote, "Common sense is the collection
of prejudices acquired by age eighteen,"[7] he obviously was referring to
prejudices, customs, or conventional wisdom rather than Reid's philosophical concept of common sense. Einstein also wrote, "The whole of
science is nothing more than a refinement of everyday thinking."[8]
Science can be used to clarify and refine our understanding, but not to
replace our common sense.

Reid argued that common sense, meaning the natural judgment of
common people, is the ultimate judge of reality.[9] He argued that reality
or true knowledge can be found only in human consciousness and
human understanding of reality, and thus neither needs to be proven nor
can be proven because human intellect and understanding ultimately
must also provide the grounds for all proof. Other Scottish philosophers,
including Thomas Brook, William Hamilton, and James Mackintosh added
refinements to Reid's philosophy of common sense and extended it to
deal with direct knowledge of human *morality* as well as *reality*.
According to this eighteenth-century philosophy, common sense is our
inner sense of first principles, by which people must test both the truth
of knowledge and the morality of actions.

For the past four hundred years, people have been encouraged to
suppress their spirituality, to rely instead on scientific reasoning and
rationality. They have been encouraged to abandon, or at least ignore,
their belief in the value of common sense. It has been called the "wisdom" of the foolish, the uneducated, and the unscientific. But, spirituality has survived this derision and so has common sense. Surveys, taken
all over the world, indicate that nearly all people believe in some form
of spirituality – the existence of some higher power or higher order of
things that is accessible to all people. The belief in common sense is still
pretty common.

Conventional wisdom, on the other hand, is fundamentally different
from common sense, although the two are sometimes mistakenly used

interchangeably. Both represent widely held opinions, but the sources of those opinions are quite different. Conventional wisdom is rooted in the science of logic and reason – in conclusions drawn from past observations – although the logic and reasoning frequently are faulty, and thus the conclusions also are faulty. Conventional wisdom need not be based on first-hand observation, instead being passed down from generation to generation. And, conventional wisdom may include some things that make common sense. However, something *makes sense* to us only if we *sense* it is true – only when the truth of it is validated by the spiritual or metaphysical rather than the physical or thinking part of our being. Some people choose to deny their common sense, and instead rely solely on scientific logic and reason. But, we all have access to common sense, if we choose to use it.

Even the Founding Fathers of the United States were capable, at times, of denying their common sense in favor of the conventional wisdom. The rightness of owning slaves, for example, was conventional wisdom until well into the 19th century – it had always been done. However, it has never made common sense that one person should enslave another. Thomas Jefferson wrote and spoke out against slavery, because he knew it was ethically and morally wrong. Yet, he helped draft a constitution that allowed slavery, and he personally owned slaves. Jefferson allowed conventional wisdom to take precedent over his common sense.

Until the 20th century, women in the U.S. were denied the right to vote; by conventional wisdom, their husbands should vote for them. In fact, former slaves were given voting rights in the U.S. before voting rights were granted to women. It didn't make common sense to deny women their voting rights, not now and not then. Thomas Paine, among other prominent revolutionary leaders, spoke out in favor of women's suffrage in the writing of the U.S. Constitution. But again, the leaders of the country allowed conventional wisdom to take precedent over common sense.

It was once conventional wisdom that war was a legitimate means of acquiring new territory, just as slavery was a legitimate means of controlling other people. For the most part, that conventional wisdom has been discarded, although still frequently violated by those unwilling to abide by common sense. Today it is conventional wisdom that purchasing is a legitimate means of acquiring virtually anything one can afford and that employment is a legitimate means of controlling other peoples' lives. As humanity has continued to progress, as humans have learned and evolved in their thinking and understanding, much of the conventional wisdom of the past has become today's foolishness. It's inevitable that some conventional wisdom of today will also become tomorrow's foolishness.As humanity progresses in its thinking, our conventional wisdom will change, but not our common sense.

Our common sense never changes. Common sense has always told people that war has never been a legitimate means of acquiring anything and that slavery has always been morally wrong. Today, our common sense tells us that no one necessarily deserves anything just because he or she has enough money to buy it, and that no one has the right to control the lives of other people just because they employ them. Our common sense also tells us that the human mind is incapable of fathoming the limits of creation or discerning truth and reality with certainty. Conventional wisdom will eventually change, but common sense has not, does not, and will not.

Although we sometimes confuse conventional wisdom for common sense, we need only stop and think for a moment to discern between the two. Do I *believe* this to be true because of what someone has told me, because of something I have read, or something I have reasoned through on my own – or perhaps, because I simply want it to true? Or do I *know* this is true because I feel it in my heart and soul? If it is common sense, you will know that it is true because you feel it in your soul, through intellectual insight rather than empirical observation or reason. Otherwise, it's only conventional wisdom.

The framers of the American Declaration of Independence had no scientific basis for the bold assertion:"*We hold these truths to be self-evident, that all men are created equal, that they are endowed by their Creator with certain unalienable Rights, that among these are Life, Liberty, and the pursuit of Happiness.*" These truths were not derived by logic and reason, and this statement certainly did not represent conventional wisdom in those days. But they felt the truth of it in their souls. They were relying on their common sense.

There is no logical, rational reason to support the Golden Rule, "*Do unto others, as you would have them do unto you.*" Yet, this powerful concept is a part of almost every organized religion and every enduring philosophy in the history of the world. It's just common sense. When Paine and others before him wrote of "the rights of man," and others since have written about "basic human rights," they did not rely on exhaustive scientific experiments. They relied instead on common sense. Our common sense, like theirs, comes from the spiritual part of us that allows us to glimpse the realm of the higher order of things. We all have access to it, but we must open our hearts and our minds to receive it. We must be willing to accept its reality.

So how does all of this relate to my near-death experience and to my purpose for still being alive? First, I'm convinced that most people, as I was, are going through life without taking the time to stop and think, or to ask *why*. Why do we live where we live, marry or stay single, stay married or get divorced, have kids or not have kids, wear the clothes we wear, drive the cars we drive, and live in the houses we call home? Why do we work at our particular jobs, work the long hours we work, and do the things we do to get ahead? What is the purpose of our life as a whole, which in turn must give purpose to the individual things we do?

Restoring a sense of purpose to our lives will require a revolution in our way of thinking. It will require a return to our common sense of good and bad and of right and wrong to provide a moral and ethical foundation for a new view of the world and of our place within it. We are at a point in time, not unlike the period of "enlightenment" of the 1700s, when the old religious era was dying and the "age of reason" was struggling to be born. Except this time, the old era is the "age of reason," rather than of religion. The new era might be referred to as the "age of insight" – an age of intelligent insight and thoughtful intuition, rather than mechanistic scientific inquiry. In the age of insight, we need not reject science, but we must accept that science can only tell us *how*, and not *why*. Intellect combined with insight can be powerful, but intellect without insight is dangerously potent. In the age of reason, we looked to science to discover knowledge. In the age of insight, we must look to the spiritual to rediscover purpose and meaning.

I am not suggesting a return to religion. Spirituality is not synonymous with religion. In fact, organized religion has become almost as decadent as the rest of society in that it has abandoned the spiritual in deference to the rational. The age of insight will bring a return to true spirituality – spirituality in the sense of seeking harmony with the higher order of things. We will rely on spirituality to find meaning and purpose for our lives and to define the guiding principles for purposeful living. But, we must reject any *new religions* that propose new sets of laws, rules, and regulations for day-to-day living. For such day-to-day matters, we can still rely on science and rationality, as long as our science and rationality is tempered with common sense. The age of insight will bring us something completely new. We will see the world anew, from a new and different perspective. Only our insights and intuition are capable of guiding us through the great transformation to a fundamentally new and better world. But first, we must break away from the old; we must experience a revolution of new ideals.

In the writings of Thomas Paine, I began to discern the anatomy of such a revolution. First, Paine gave the enemy no quarter. From Paine's writings, you would have thought that Great Britain was hell and King George was the devil himself.[10] Paine denied the legitimacy of the

Monarchy. He said those of past generations, who had signed away for all times their rights to self-rule, had attempted to give away rights that they did not own and therefore could not legally sign away. The crown was persecuting the colonies mercilessly, he said, because the concept of Monarchy was inherently oppressive. Paine concluded: the only way to end the tyranny was to declare our independence from Great Britain.

However, Paine always went beyond condemnation in his writings. He always painted a vision of the great and glorious nation that lay in the future, just beyond independence. He met every argument concerning what the colonies would lose with an argument for what they would gain. They would lose the protection of the British fleet, but as Paine pointed out, the only enemies the Colonies had were actually enemies of Great Britain. The trade they obviously would lose with Great Britain would be more than offset by new opportunities for trade with other countries – trade with the enemies of Great Britain. However, most important, the freedoms of a democratic republic would eliminate control by the Monarchy, which would allow the colonists to shape their own destiny. Democracy would restore to them the basic "rights of man" which the Monarchy had taken away. All "men" long for freedom, he said, and with an independent republic, they would be free.

Finally, Paine never gave a hint of doubt that the revolution would succeed. Even it its darkest hours Paine found common sense reasons to support his optimism. For example, when the British occupied Philadelphia, some saw it as a major defeat for the colonies. However, Paine saw it as an indication of the inevitability of victory. If half of the British army was required to occupy one city, they could never possibly expect to occupy all of the vast space that constituted the whole of the American colonies. Also, when the British took the fort at Charleston, Paine called it a blessing in disguise. The Colonials had become complacent and lethargic in carrying out the war. Now they would be compelled to take notice, to get serious about the war, and to move boldly toward final victory. Paine never doubted the ultimate victory – to him, it just made common sense.

So that's why I had checked out Thomas Paine's book from the library before going in for my stress test that day. I needed to learn the secrets of successful revolution. We must identify the villain and know why we are revolting, we must have a clear, positive vision of what we expect the revolution to achieve, and finally, we must believe in our hearts that the new revolution of ideals can succeed. That's why my heart stopped for a minute or so that day in February of 1998. I needed to have time to become acquainted with Thomas Paine. But most importantly, I needed to have time to stop and think.

In my days and weeks of recuperation, I came to realize that it is time for a new kind of revolution in America. I had been reading about the post-industrial, knowledge-based society, about quantum physics as the new foundation for science, about growing skepticism of science and technology, and about returning to spirituality to discover basic principles for living. I had also been thinking about the inadequacy of economics as a social science, and about the extent to which economics now dictates so many aspects of our lives. And, I had been working professionally on issues related to sustainable development, with an emphasis on sustainable agriculture. But, I had not yet put all of these things together in my mind. Now, I began to understand the nature of the new villain – a materialistic, corporatist society. But, I also had a new vision for a better future – a holistic, sustainable society. And I could see that the victory would be not only possible, but also inevitable, if we could only find the courage to challenge the current conventional wisdom with our common sense.

My brush with death gave me both the time and motivation to realize the need for revolution, but it also helped me realize that the purpose for the rest of my life was to promote this new revolution of ideals. I was not to be a revolutionary in the sense of Thomas Paine, supporting and promoting revolution through armed conflict. This new revolution would not be fought on the battlefields, or even in the halls of Congress. Instead, it would be a battle for the hearts, minds, and souls of people. My purpose was not to call people to arms, but rather to challenge people to stop and think – to ask why. My purpose was to help people find the courage to challenge the pseudo-wisdom of science and rationality – to learn to rely instead on their intelligent insight, on their common sense.

My years of academic and professional experience have given me unique credentials to help people understand the abuses of science and reason, and to help them throw off the tyranny of the economics of greed. I also have gained some unique insights into how to build a more sustainable society – a society that is socially just and ecologically sound as well as economically viable. And I had reason to believe that such a revolution could be successful, in fact would be destined to succeed, once people realized it would lead to a more desirable quality of life. But the success of this new revolution depends upon the people, upon the willingness of people to rely on their intelligent insight – their common sense. My primary purpose for being, for the rest of my life, is to help make the case for *a return to common sense*.

Endnotes

[1] Thomas Paine, *The Life and Major Works of Thomas Paine*, edited by Philip S. Foner (New York: The Citadel Press, 1961; republished, 2000 by Replica Press, Bridgewater, NJ).

[2] Thomas Paine, *Common Sense* (Mineola, NY: Dover Publications, 1776, republished 1997).

[3] Hugh Elliott, "Materialism," in *Readings in Philosophy*, ed. John Randall Jr., Justus Buchler, and Evelyn Shirk (New York: Harper & Row, Publishers, 1972): 309-310.

[4] *Microsoft Encarta Encyclopedia*, 2003, "Plato: Forms" (Redmond, WA: Microsoft Corp., 1993-2003).

[5] Thomas Reid, *Works of Thomas Reid*, ed. William Hamilton, Thoemmes (Bristol, England: Continuum Press, 1863), 422.

[6] Thomas Huxley, *On A Piece Of Chalk* (New York: Scribners, 1967, 1st Edition, 1869).

[7] Albert Einstein, *Mathematics, Queen, and Servant of the Sciences*, quoted by E. T. Bell (Washington, DC: Math Association of America, 1987, original copyright, 1937).

[8] *Chemistry Coach*, "Common Sense, Albert Einstein," Bob Jacobs, <http://www.chemistrycoach.com/common_sense.htm> (accessed September 2006).

[9] *Catholic Encyclopedia*, "Philosophy of Common Sense," New Advent, <http://www.newadvent.org/cathen/04167a.htm> (accessed September 2006).

[10] Paine, *Life and Works*

The Villain

2

GLORIFICATION OF GREED:
THE PERSONAL VILLAIN

As I recovered from surgery, making my way back into the real world, I began to realize the enormity of the task before me. I was ready for a revolution in thinking, but most of America seemed to be rushing blissfully toward what I saw as an abyss of self-destruction. In the spring of 1998, the American economy seemed strong, but the strength was an illusion propped up by years of denial. I thought back over the years to the events that had kept moving the nation toward the brink. The denial began during the 1970s, when the government attempted to use inflation to ease the burden of rising energy costs. If people had more money to spend, then gas prices would not seem so high. But inflation got out of control. We ended up with low growth accompanied by rising prices, rising interest rates, and rising unemployment – stagflation, it was called.

We eventually escaped from the grasp of chronic inflation during the early 1980s with monetary discipline by the Federal Reserve and free market conservatism of Reaganomics. But the country didn't escape without a severe economic recession. Workers were told they must make major concessions to help American industry weather the recession. Corporate downsizing and outsourcing caused millions of industrial workers to lose their jobs. Eventually the labor unions were broken as people became "really scared" about their future. Many were willing to work almost anywhere, under any conditions, for whatever wages the corporations were willing to pay. Most new jobs were lower paying service jobs and many were part-time jobs without benefits, but people eventually found work, in unprecedented numbers, and unemployment fell like a rock.

The battle cry of the Reagan era had been, "get the government off our backs," and let the market work. Congress had responded by dismantling or disabling many environmental programs to remove "undue interference with corporate profits and economic growth." During the 1970s, we had succeeded in making our lakes and rivers non-flammable again and in bringing daylight back to the big cities, but now we were told we had taken environmentalism too far. A strong economy was more important than a clean environment, at least to those in positions of power and influence.

The government had also reined in its antitrust lawyers. American corporations needed be able to merge so they could become big enough to compete with government-supported industries of other countries. Not only were there economies-of-scale in production and distribution, but larger corporations could afford to invest more in research and development, to bring more new products and services to the market place. Increasing public research and development would have required more tax dollars and the people were being told that virtually all government spending was wasteful. So development of new technologies, including medical and biological technologies, for the most part was left to the private sector – to large, corporate firms. The Department of Justice rejected very few proposed corporate mergers during the '80s and '90s. Still fewer cases of non-competitive market behavior were pursued in the anti-trust courts – most being high profile cases, seemingly chosen to create the illusion of enforcement.

The George H.W. Bush and Bill Clinton Administrations had only restated and refined the trends set in place during the Reagan years. Whereas the clear message of the Reagan era had been that *greed is good*, Bush had attempted to take some of the harshness out of the rhetoric by calling for a *kinder and gentler nation*. Clinton had put forth a more liberal political agenda, but he never seemed willing to give priority to principles over politics. The rallying cry of his campaign had been "*It's the economy, stupid*," and apparently he never let anything take precedent over corporate profits and economic growth. The economy boomed and the stock market soared. The huge federal budget deficits, carried over from the Reagan years, were suddenly transformed into record budget surpluses. Previous arguments over where to cut the federal budget turned to debates of whether to cut taxes or fund new social programs.

A new era of informational and biological technology also was arriving, in full stride. New information enterprises began creating new wealth and creating new high-paying jobs at unprecedented rates. A young fellow by the name of Bill Gates, created a natural monopoly called Microsoft. His "disk operating system" (MS-DOS) became the industry standard, allowing different brands of computers to communicate, and within a couple of decades, he would become the richest man in the world. The Internet followed, bringing a flood of enterprising young entrepreneurs with companies such as Yahoo, E-bay, and

Amazon.com, whose stock prices soared before they had ever reported a dime in profits. Thirty-year-old millionaires became a dime a dozen in Silicone Valley and other places that had staked their economic future on new information technologies.

Genetic engineering had yet to approach the information technologies in creating profits and economic growth. Their promise was still on the horizon. However, a company named Pfizer developed a drug called Viagra, the ultimate high-tech aphrodisiac. They hyped it on network news shows, hired Bob Dole to hawk it in TV ads, and the Pfizer stock price soared – setting the economic standard for drugs of the future. Erectile dysfunction is a real medical problem for some, but those in real need were not the only viewers of this massive national advertising campaign. A lot of basically healthy men also wanted an enhanced erectile function, and were willing to pay a lot of money to make it happen.

In addition, corporations were extracting billions, perhaps trillions, of dollars in profits and growth from the natural environment, from workers, from consumers, and from taxpayers. Deregulation, corporate tax credits, and outright government subsidies "bought" through political influence all facilitated the process. This process of extraction and exploitation became worldwide as new agreements on international trade began paving the way for a single, global market. Some of the economic growth was real, arising from new post-industrial technologies. But the real growth had been commingled with growth extracted from society and the environment, making it difficult to separate real productivity from exploitation.

Stock markets cannot distinguish between speculative fantasy and economic reality. With a sustained 20 percent annual growth rate during most of the 1990s, the U.S. stock market boomed, breaking through the "10,000 ceiling" on the Dow Jones stock index. Alan Greenspan, Chairman of the Federal Reserve Board, at one point warned the nation about "over-exuberance and unwarranted growth" in stock prices. The market faltered only briefly, and then soared upward again. Greenspan later started talking about the "new economy" and eventually gave up trying to explain why stocks continued to rise. There simply weren't any rational explanations.

How much of the economic boom of the '80s and '90s was real and how much was illusion? How much of the growth in corporate profits represented increases in real productivity, and how much of it represented corporate exploitation? How much of the economic recovery since the recession of 2001 is due to huge government deficits and how much is due to growth in productivity? Such things are virtually impossible to disentangle, even after the fact.

All rising markets eventually fall, so the critical question is not whether recessions follow expansions. They do. And expansions follow recessions. The critical question is whether the long run trend in economic growth can be sustained. As we continue to go through cycles, will future cycles take the economy to higher or lower levels? Growth based on exploitation, extraction, speculation, and illusion – rather than real productivity – quite simply is not sustainable.

The new revolution will not be won until people listen to their inner insight – their common sense – and acknowledge that our current system of economic development is not sustainable. Only when they admit to themselves that their overall quality of life is headed down, rather than up, will people demand fundamental change. As long as they can rationalize their continuing confidence in the prevailing economic and political philosophy, people will not willingly support real change. It is difficult for an addict to admit the need for help while he or she still has plenty of drugs on hand. So, most people, at least those with marketable skills and talents, will cling to the hope that if they continue to work hard enough, long enough, ultimately they will achieve success.

For many others, work has become its own reward. They continue to move from task to task, from deadline to deadline, from success to success, and with each success, to another high. But each high is followed by feelings of emptiness. Too few, however, stop and think about whether the highs and lows have any real meaning or whether their lives even make sense.

The task before us is truly daunting, but all revolutions start small. Revolutions actually begin when only a few people realize that something is critically wrong, that it can't be "fixed" by simply fine-tuning, and the only real solution is fundamental change. These early revolutionaries debate the issues, refine the principles, and test new strategies. This is the way new paradigms typically are created. These new alternatives are thus already available when the old regime begins to crumble and people finally find the courage to rebel. The American Revolution did not begin with the Declaration of Independence in 1776 – it had been growing among a handful of colonists who had been working for decades toward a new paradigm of government, a representative democracy. The new revolution may be starting out small, but it is nonetheless underway. We don't need to change the world all at once; we just need to change it a little bit each day. We don't need to change the whole world all by ourselves; we just need to change our little part of the world. The world is an interconnected whole. As we change our little part, day by day, we in fact are changing the world. We are moving the world toward a great social, economic, and ecological revolution.

Economies are creations of societies and societies are made up of people. And people themselves must be responsible for ensuring that their economy serves their interest. Thus, if something is fundamentally wrong with the American economy, something also is wrong with the society that created it and has responsibility for it, meaning there is something fundamentally wrong with us – the American people.

Revolutions begin by identifying the villains – the things that must be changed – beginning with the villains within. I am convinced the personal villains of most Americans today are selfishness and greed. People inherently pursue their self-interests; it's just human nature and it's not inherently wrong. But selfishness and greed go beyond normal self-interests, and historically these were labeled as socially and morally unacceptable. So how did America, with its roots in democracy and religious freedom, become a nation that so willingly surrenders to the seduction of selfishness and so openly proclaims the glory of greed? By the late 1990s, I was beginning to understand the answer to the *how* of this question, if not *why*.

My lessons in economics began early in life. I had grown up on a small dairy farm in southwest Missouri. When I was young, we were poor. Many people say proudly, "We were poor, but we didn't know it." I knew it. I have never liked getting into "poor contests" because I feel sorry for myself when I win and feel even sorrier for the other person when I lose. So, I will elaborate no further on my childhood poverty.

My mother always told me that she wanted me to "amount to something." I took that to mean that she wanted me to have more money, more prestige, or more of something that we didn't have while I was growing up on the farm. My dad always said that I should get a good education, because that was something "they can't take away from you." I didn't understand who "they" were or why they would want to "take something away from me." But my dad had bought our farm during the Great Depression and had nearly killed himself paying for it, so nobody would be able to take it away from him. As I would understand later, there are still people out there who live by taking things away from other people. Mostly, my dad just didn't want me to have to work as hard as he had worked for a living.

When I graduated from high school, I had made pretty decent grades and my parents encouraged me to go to college. With a $125 Curators Scholarship from the University of Missouri, we scraped together enough money to get me through one semester. This would give me time to find a job and then save enough for the second semester, and to continue working my way through the next three years. I worked hard, and our plan worked well.

When I started to college, I didn't know what I wanted to study; anything that would prepare me for a good job would do. I had never heard of Agricultural Economics when my advisor enrolled me in Ag Econ 100. However, it didn't take long for me to decide that Agricultural Economics was "my calling." Economics was about making money, and that was exactly what I wanted to learn how to do.

I graduated from college in the spring of 1961. When I walked across the stage to get my diploma, I had the promise of a good job in the meat packing industry. My dreams and aspirations at that time probably were pretty much the same as those of millions of other young men all across the country who were at that same stage of life. We wanted a good job with a good company, so we could live a reasonably affluent lifestyle, maybe even get rich, then retire, and in general, have a good life. We wanted to live the American Dream.

My dad had taught me that my reputation would be my most valuable asset in life, so I should be honest and never try to take advantage of anyone else. My mother had taught me that I shouldn't think of myself as any better or worse than anyone else is, and I should treat other people, as I would like to be treated. I was raised to believe that selfishness and greed were basic human frailties, if not outright sins.

But four years of economics and business classes erased much of this "agrarian" childhood training. At graduation, I had few remaining reservations concerning the righteousness of selfishness and the goodness of greed. I had learned that the way we can do the most good for society as a whole is to do the most we can for ourselves. At first it didn't seem quite right, but after some thought, I could see that it was all perfectly logical and rational. In fact, through economics, I could see that if I failed to pursue my self-interests to the fullest, I would not be doing my part to help build a strong society. Economics is a powerful discipline. It shows us, logically and rationally, that we do not have to conquer our basic selfish instincts, but instead we need only cultivate them. I now understand the fallacies of my conclusions. At the time, however, I didn't have enough real world experience to challenge what I had been taught.

I realize that many people who read this book will have never taken a course in economics, many others will have had a course in economics but have learned little if anything, and even those who think they learned something may not have found it very useful or relevant to their day-to-day life. Courses in economics can be pretty dry and boring, with all of the charts, graphs, and tables of statistics. As a result, the economic literacy level in the U.S. is pretty low. This widespread lack of economic literacy leaves members of the general public unable to refute or even to debate the illogical economic arguments put forth in support of selfishness and greed. Thus, people grudgingly accept statements such as; if it's profitable, it's inevitable; if it's not profitable, no one will do it; it's the economy, stupid, that's all that matters; a rising tide lifts all boats; and the pursuit of individual self-interests is good for society.

We all know these things don't make sense in our day-to-day world, but most people just don't understand enough about economics to understand or explain *why*. Those of us who have had courses in economics probably have never had an instructor explain to us that we can use our knowledge of economics to confront the economic fallacies that bombard us every day. I know none of my instructors did. But I have since learned that knowledge of *basic* economics, just the most basic of economic concepts, can be powerful defense against the economic dogma that permeates our society today. The most powerful economic concepts also are the most basic and simple to understand. It just takes getting over the fear of learning about something that challenges our understanding of reality. It's true that many people become seduced by economics, because it seems to justify their greed. But knowledge of economics can also provide a powerful defense against its seductiveness.

The logic of economics isn't all that difficult. We're not talking about rocket science here, just the realities of everyday living. The next 1200 words explain all of the important economic concepts you need to understand to make you more economically literate than 99% of American adults today. With this basic understanding, you will know why the economic charlatans take the positions they take and will be able to expose their fallacies. Economics really is mostly a matter of supply and demand. The price that people are willing to pay for something reflects its economic value. If something is high priced, it has a lot of economic value, at least to someone. If its price is low, it's not worth much economically to anyone. Economic value obviously is different from other kinds of value. The economic value of a thing is related to its scarcity – how much of it there is relative to how much people want. Something like air obviously is valuable – people have to have it to live. But air has no economic value, at least not as long as there is plenty for people to breathe all they want. But, air does take on an economic value when we pollute, making *breathable* air scarce. Diamonds certainly are not necessities of life, but they have a lot of economic value. There aren't many diamonds around, so those who really want them have to pay dearly for them.

The ideas of scarcity and value provide the foundation for the laws of supply and demand, both of which are derived from an economic concept called "diminishing marginal value." On the demand side, value refers to the value of something to consumers or buyers. The word "marginal," in economics, refers to adding things incrementally – one after another. So "diminishing marginal value" means the value of a thing diminishes as additional amounts of it become available. In other words, each incremental unit of something becomes less economically valuable than the unit before.

The demand for Big Macs is an example used in Economics 101 these days. At any given sitting, the first Big Mac might be worth two bucks, but the second one would be worth something considerably less, at least for most of us. After the third or fourth Big Mac, the marginal value of another would drop sharply, and eventually would drop to nothing, even for a hungry teenager. Whether it's Big Macs, Nikes, or Hondas, as something becomes more available, it becomes less scarce and its economic value declines. Conversely, if less of something is available, it is more scarce and more economically valuable.

A market is nothing more than a collection of individual buyers and suppliers. Thus, the "law of demand" states that if the quantity of anything offered for sale *increases*, its market price will *decline*. On the other hand, if the quantity of something *declines*, its market price will *increase*. Thus, we have the law of demand: "Quantity demanded and price tend to move in <u>opposite</u> directions." If you think about it, it just pretty much common sense.

The supply side of the market is a bit more complex, but the basic idea is much the same. For producers of the things people buy, if they increase amounts of production inputs used, such as labor, land, capital, or raw materials, they generally see an increase in their output or production. But as use of these inputs are increased, beyond some point, an additional increment of a given input will not add as much to total production as was added by the previous increments. As the "marginal productivity" of the additional input declines, it takes more inputs to produce a given amount of output, and the "marginal cost" of production increases. Producers can afford to buy additional inputs to increase production only so long as they expect to get a price high enough to cover their increasing costs of production. If prices are expected to go up, producers tend to produce more. If they expect prices to fall, they tend to cut back on production.

As with market demand, market supply is nothing more than the total of quantities offered for sale by individual producers. As market prices *increase*, quantities supplied will *increase*, and as market prices *decline*, quantities supplied will *decline*. This is the essence of the law of supply; "Quantity supplied and prices move in the <u>same</u> direction." Again, this is not rocket science, it just makes common sense.

The laws of supply and demand are all we need to understand the magic of markets. In fact, that is about all there is to economics. As John

Kenneth Galbraith, a noted twentieth-century economist once wrote, "The great truisms of economics have no clear discoverers; they are evident for all to see."[1] They are just common sense.

Quantities demanded and quantities supplied tend to move in opposite directions, relative to market price. As prices go *down*, quantities demanded *increase*, but quantities supplied *decline*. As prices go *up*, quantities demanded *decline*, but quantities supplied *increase*. In spite of misstatements to the contrary, this ensures that the quantity supplied is always equal to the quantity demanded – the market always clears. At any price above the market-clearing price, the quantity demanded would be less than the quantity supplied, but prices would then fall. At any price below the market-clearing price, the quantity demanded would be greater than the quantity supplied, but prices would then rise. Producers' expectations of prices are sometimes incorrect, and they produce either more or less than will clear the market at their *expected* price. However, falling prices will eliminate any temporary surpluses. Likewise, rising prices will eliminate temporary shortages. Demand never exceeds or falls short of supply; the markets always clear.

Markets also determine wage rates for labor, interest rates for capital, rental rates for land and other real estate, and salaries for managers. The demand for inputs or factors of production is derived directly from prices and profits at the consumer level. Higher prices and more profits from a given product result in greater demand and higher prices for the things it takes to produce it. Falling product prices translate into weaker prices for production inputs.

The supply of production inputs responds to prices in the same way as the supply of consumer products. If wage rates fall in a given industry, workers move to other better paying jobs elsewhere. As producers in a growing industry attempt to expand production, they need more capital, they bid up interest rates, and capital flows in from other uses to finance their expansion. Thus, productive resources – land, labor, capital, and management – are shifted from one use to another as changing profitability make them more or less valuable in producing different goods and services. They are shifted from less valued to more valued economic uses. It just makes sense that things are worth more if they can be used to produce other things that are worth more, and things are worth less if the things they are used to produce are worth less. Again, it's pretty much a matter of common sense.

With this basic understanding of economics, a person can understand how a free market economy is supposed to work. This is the essence of *microeconomic* theory – the theory of behavior of the individual consumers and individual business firms. Individual consumers seek to maximize their utility or satisfaction and business firms seek to maximize profits. As each pursues their individual self-interest, they are guided, as if by an invisible hand, to allocate scarce resources so as to maximize the collective economic welfare of society.

So how has microeconomic theory promoted the causes of greed and selfishness? First, in our role as consumers, economics tells us we should never pay more than necessary to get whatever we want. Not only would doing so unduly reward producers, but it would distort the market signals causing production of the thing to increase beyond levels justified by its economic value. Likewise, we should never outright give money to anyone. We would be rewarding them, even though they did nothing of economic value. Economists justify charity by saying that we are actually buying some sort of personal satisfaction from the act of giving. But, they are still concerned that charity rewards someone for their non-productivity and will only encourage them to be even less productive in the future.

To economists, it is irrational to spend money on anything for which there is no individual, personal reward, and doing such things messes up the economy. We can work collectively, through government, to take care of such non-economic issues as social equity and resource stewardship. But in spending our own money, we need to practice absolute selfishness, if we are to do our part for the overall good of the economy and society.

As owners of productive resources, we should do the same thing; the pursuit of individual greed results in the greatest good for society as a whole. We need to work in the occupation where we can earn the highest possible wages or salary. If we continue to work at some lower paying occupation, like farming, teaching, or police work, because it's a family tradition or just because we like it, we are messing up the labor market. We are restricting the labor supply in some other higher valued occupation that is producing things that people want more. If we could not be more productive elsewhere, we would not be able to earn a higher salary elsewhere. A good economic citizen is obligated to work for as much money as they can get.

Likewise, if we have money saved, we should invest it where it will yield the highest return. That's where it will do the most good for the economy. If we sock it away under our mattress, build up a balance in our checking account, or God forbid, give it to our kids to help them buy a house, we are really fouling up the capital markets. We are restricting the availability of capital elsewhere, where it could be used more productively, and are forcing interest rates higher than necessary for those who are trying to produce things consumers really want. If you insist on farming land that would be worth ten times as much for residential development, you must be just plain nuts. If you aren't earning as much as you can from everything you own, then you are not really doing your part for the good of the economy.

The same reasoning holds for the large corporation or the individual entrepreneur. Individuals and corporations should invest their resources in the enterprises that promise the most profit and potential for growth. The prospect of profits suggests that consumers want more of something than current producers are willing to provide. By taking that profit, you are simply responding to market demand. If you could supply a more profitable market and don't do it, you are depriving people of an opportunity to consume more of what they want. If you are not making as much profit as you can, you are not doing your part for the overall economy.

In economics, by definition, human wants are insatiable and resources are scarce. Thus, we can never have "enough." The more we get from our scarce resources, however, the better off we will be. So the goal of economics is to make as much money as possible so the economy will grow as fast as possible and become as large as possible. Economics is a study of the allocation of scarce resources among competing ends. The more efficiently we allocate resources, including people, among competing uses, the larger our economy can be, the faster it can grow, and the wealthier we can become. No other means can allocate economic resources more efficiently than can free markets, but only if everyone maximizes their individual economic self-interest.

The greatest greed results in the greatest good; it's rational, logical economics. We don't have to feel guilty about our human frailties. The more we want, and the more we are able to get for ourselves, the more opportunities we create for everyone else.

Virtually every person in America has been indoctrinated with this economic doctrine. Many people have been indoctrinated directly, through one or more economics courses in high school or college. Regardless of whether they understood the theory, they remember the seemingly well-reasoned, logical conclusions: by pursuing our self-interests, we are serving society. Those who haven't had a course in economics were indoctrinated by instructors in some other courses, such as history, government, and sociology, who had acquired a faith in the ideas of "laissez faire" economics. Those who haven't been exposed to the doctrine of free-market economics in the classroom are bombarded with it daily – on television, in newspapers, by politicians, by business leaders, by their neighbors. The relentless message is that competition, free enterprise, self-reliance – the pursuit of self-interest – "is what made America great." The economics of individual, self-interest has become our "national religion." In truth, it has become the national villain which blinds most Americans today.

With a degree in Agricultural Economics in hand, I thought I could easily cope with any twinge of conscience I might have left about being greedy and selfish. I was ready to do my part for society, by making as much money as I could. However, somewhere in the back of my mind lingered questions that would not go away.

What was the "true value" of something? Somehow, basing the value of everything on its economic value seemed a bit inadequate; economics only valued things that were scarce. A world of insatiable wants, where there was never enough, did not seem to me to be a very hospitable and peaceful place to live. Was it really the nature of humans to express their selfishness and greed, or was that just the nature of the animal within us?

How could all of the religions of all times be so wrong? How could the various forms of government be so wrong? Was the sole economic role of the government and society just to "let the markets work?" How could the pursuit of selfishness and greed be the keys to creating a more desirable society?

Somehow, a lot of what I learned about economics just did not seem to make sense. Even with these earlier lingering doubts, however, it would take many years to gain the courage to rethink what I had learned. Only later in life would I come to realize that selfishness and greed have become the villains in America – the personal villain within, against which we must revolt.

Endnote

[1] John Kenneth Galbraith, *Economics in Perspective: A Critical History* (New York: Houghton Mifflin Company, 1987).

3

INDUSTRIALIZATION OF AMERICA:
THE ORGANIZATIONAL VILLAIN

During my senior year in college, I interviewed with Wilson & Co., one of the largest livestock slaughter and meat processing companies in the country at that time. Swift, Armour, Wilson, and Cudahy had been known as the "big four" of meat packing, although some new players, including Oscar Mayer and Iowa Beef Packers (IBP), had come on the scene. When Wilson made me an offer of employment, I accepted. I had to do a six-month stint in the Army Reserves before I could start work, and my military duty was delayed by the Berlin crisis of 1961, but in the fall 1962, I reported for work at Wilson & Co. in Kansas City, Kansas.

While I didn't realize it at the time, I had just become a part of "American industry." I was taking my place in a company that epitomized the industrial model of production and distribution. A packing plant might be more accurately described as a disassembly line rather than assembly line, but the basic characteristics are the same. The animals come in whole and the workers take their carcasses apart, piece by piece as they move down the disassembly line. Only far later would I realize that the industrial model dominates nearly every aspect of the contemporary American economy and society, including not just business organizations but also education, research, national defense, volunteer, and non-profit organizations. Even our religious institutions have become industrialized. The "age of reason" has reached its pinnacle in these latter stages of the "era of industrialization." Specialization of function, standardization of process, and consolidation of control; these are the basic means by which we organize our activities and carry out virtually everything we do.

Most people probably haven't stopped to think about it, but the organizational ideas of industrialization and capitalism have come into vogue only over the past couple of hundred years – only recently, by historical standards. During medieval times, people had completely different ideas about how business ought to be organized and how things in general ought to be done.

Before the industrial era, most independent merchants and crafts-people were organized into local guilds.[1] Some guilds were formed for social and religious purposes, performing many of the functions of today's civic organizations and local governments. Merchant guilds were formed to control local markets, regulating local pricing, quality standards, and terms to trade for specific commodities. Craft guilds were producer organizations with members classified according to the products they produced, including weavers, masons, and carpenters. In general, guilds were a combination of local government, trade association, local monopoly, and consumer protection associations. Guilds ensured that all within the community were treated equitably, if not always well. But many local guilds were openly hostile to outsiders. Their rules of trade and methods of operation were more strongly influenced by local customs and religious values than by principles of economic efficiency.

The industrial revolution began in the late 1700s with development of the British factory system for mass-producing textiles.[2] Inventions patented during a span of two decades fundamentally transformed the future of human society, beginning with John Key's "spinning jenny" in 1764 and ending with Edmund Cartwright's "power loom" in 1785. Although the industrial revolution was marked by mechanical inventions, these provided only the means by which the mechanistic concepts of science and organization, which had been around for a couple of hundred years, could be effectively applied in production and commerce.

In 1769, James Watt, a Scotsman, patented the steam engine, providing a power source for British textile mills during the late-eighteenth century and for nearly all of nineteenth century industry everywhere. Eli Whitney, an American inventor, provided a missing link for the textile industry in 1793, when he invented the cotton gin. The cotton gin brought a fifty-fold increase in efficiency of labor in removing seed from cotton bolls, opening a floodgate of raw material to feed the new textile industry. The industrial revolution took a quantum leap forward.

However, it is doubtful that the technical revolution of the late 1700s could have occurred without simultaneous economic and political revolutions. In 1776, Adam Smith, a British economist, wrote his landmark book, *The Wealth of Nations*.[3] Smith's economic philosophies quickly spread around the world and provided a new conceptual foundation for economic thinking. Smith's principal economic thesis was that "division of labor" could generate large gains in efficiency and consequently could increase the wealth of individuals as well as the wealth of nations.

If a group of laborers, who were producing a given product independently, would each instead specialize in performing only one or two tasks, they could perform these specialized tasks much more efficiently. By specializing and working together, so that all tasks were performed by different people, the group of laborers might greatly increase their collective productivity. Smith used the example of several workers making straight pins.[4] He hypothesized that by specializing in cutting wire, sharpening, and forming heads, and coordinating their functions, a group of workers might increase their output by as much as forty-fold over that of each worker making pins separately. Smith's observations concerning division of labor, and potential gains from specialization, provided the conceptual blueprint for organization of factory work, for assembly lines, and in general, for industrialized mass production.

Smith also provided the foundation for the political environment needed to support the industrial revolution. He observed, "It is not from the benevolence of the butcher, the brewer, or the baker, that we expect our dinner, but from their regard to their own interest. We address ourselves, not to their humanity but to their self love, and never talk to them of our necessities but of their advantages."[5] Later, in reference to trade, Smith states, "he intends only his own gain, and he is in this, as in many other cases, led by an invisible hand to promote an end which was no part of his intention. By pursuing his own interest he frequently promotes that of the society more effectually than when he intends to promote it."[6] These statements provided the economic foundation for the industrial revolution. In a free market economy, individuals could pursue specialized individual economic self-interests and their individual pursuits would be guided to serve the public good, as if by an invisible hand.

The French Physiocrats had preceded Smith by a few years in writing about the benefits of trade and free markets. But, Smith expanded the discussion beyond gains from trade to gains from production. Perhaps more important, he put his thoughts together in a particularly compelling way at a time when the world was ready for a new way of thinking. These new beliefs, which Karl Marx would later label as "capitalism," quickly spread from Great Britain to America and Germany and later spread elsewhere around the globe.

Certainly, the industrial revolution was not completely motivated by capitalism. Some of the strongest industrial nations of the early 20th century had socialistic and communistic economies – France and Russia, for example. However, market capitalism most certainly provided the foundation for industrialization in America, and by the turn of the 21st century, had become the dominant political-economic model for industrial nations around the world.

Several fundamental beliefs define a capitalistic economy. First, capitalists believe in individual, private ownership of capital – private ownership of production facilities rather than collective ownership through government. Capitalists also believe that the best way to organize the economy is through the actions of buyers and sellers interacting in free markets rather than through central planning. Under capitalism, each consumer and producer is encouraged to pursue their individual self-interest, with a minimum of interference from government, as a means of ensuring the greatest societal benefit.

In the year *The Wealth of Nations* was published, 1776, the American Colonies declared their independence from Great Britain. It probably was no accident or oversight that the Constitution of the United States, written a few years later, makes no mention of protecting citizens from economic oppression. The writers of the Constitution apparently shared Adam Smith's belief that persons left free to pursue their own self-interests could best serve the interests of society as a whole. The U.S. Constitution stresses the right to ownership of personal property, along with protection of personal liberties, and thus, gives great attention to protection of the individual from political oppression. But it doesn't address the need to protect people from economic exploitation. Apparently, the Founding Fathers trusted the invisible hand, not only to transform individual greed into societal good but also to protect the economically weak from the economically strong. They apparently believed, along with Smith, that the best strategy for a nation to pursue in enhancing its wealth was to ensure that the government did not interfere with free and open competition among individuals, either within or among nations.

At the beginning of the industrial revolution in the late 1700s, the U.S. was an agrarian nation. The term "industrialization" is used most commonly to describe the transition from an agricultural economy to an economy based on manufacturing and trade. However, the transition from agriculture to manufacturing is but a symptom, rather than the essence, of industrialization. The prominence of specialization, standardization, and centralization, rather than a shift from agriculture to manufacturing, define the process of industrialization. The transition from agriculture to manufacturing and trade is simply a consequence of applying industrial strategies to the process of economic development.

In an agrarian economy, most people provide for most of their own basic needs and are largely self-sufficient. Self-sufficiency obviously includes provision of food, clothing, and shelter, which are the essential functions of agriculture. Specialization within agriculture increases the efficiency of agricultural production, thus creating surpluses in production, which can be traded to others who can then specialize in producing the things other than food and fiber that have come to be associated with a better way of life.

Those who manufacture things to trade with farmers, and to trade among themselves, are able to get the food, clothing, shelter, and other things they need without having to produce those things for themselves. With increased specialization, farmers become fewer, manufacturers become more numerous, and an increasing number of people are employed in producing nothing at all – they specialize in trading. Among the traders are speculators, brokers, regulators, lawyers, and a multitude of others who specialize in facilitating trade. Standardization and consolidation of control emerge and evolve as means of realizing greater efficiency from the process of specialization.

As industrialization swept across the U.S. economy during the nineteenth century, and America shifted from an agrarian to an industrial society, corporations played a major role in this economic transformation. Specialization of functions and standardization of processes allowed centralization of decision making. One manager could control far more laborers producing far greater output within an industrial organization than under the old crafts-based economy of pre-industrial days. Capital eventually became the most limiting factor to the size of industrial operations. The corporation, a concept dating back to the Middle Ages, seemed the ideal mechanism for consolidating capital from many individual investors to form ever-larger industrial organizations.

Through corporations, large numbers of investors could acquire "shares" in the ownership of a company while limiting their individual liability. For sole proprietors and partnerships, liability for debts of the company extended to all assets of the individual owners. But under corporate ownership, liability was limited to investment in the corporation. In addition, shares in corporations could be easily bought and sold, transferring ownership from investor to investor. Transferability of ownership gave the corporation an infinite life span over which to pursue profits and to accumulate more capital. To amass still more capital, two or more corporations also could consolidate assets by combining shares.

In the U.S., the consolidation of firms producing basic commodities, such as sugar, salt, leather, whiskey, kerosene, meats, and rubber goods, took place following the Civil War. This set the stage for American-style industrialization, when John D. Rockefeller formed his first corporate trust in late 1882. He persuaded stockholders in some forty different corporations involved in oil production to exchange their stock for shares in the Standard Oil Company of Ohio. This allowed Rockefeller to consolidate management and centralize decision making across a large segment of the petroleum industry under one board of directors, which he chaired. Rockefeller exerted market power over the petroleum industry, manipulating supplies, and influencing prices and profits in ways that were clearly contradictory to the conditions under which Adam Smith imagined the working of an invisible hand. American industrialists ever since that time have attempted to follow Rockefeller's lead.

By 1893, American Sugar Refining Company and the United States Rubber Company had joined Standard Oil in the merger game, before a severe downturn in economic activity brought a temporary halt to further mergers. A second flurry of mergers, beginning in the early 1900s, led to the formation of such historically well-known companies as United States Steel, Du Pont, American Can, and International Harvester.

Soon large corporations not only controlled the American economy but controlled much of the American political process as well. Politicians and elections were routinely, often openly, "bought and sold" through bribes, lobbying, and corporate financing of campaigns. Upton Sinclair's book, *The Jungle*, published in 1906 gives the reader vivid insights into the nature of early-twentieth century industrialism.[7] Sinclair documents the inhumane and unethical treatment of both animals and people by the powerful "beef trust," and illuminated its pervasive corrupting influence on the politics and government of Chicago during the early 1900s. Other trusts controlled the lives and livelihoods of other people and other cities, and together, the trusts pretty much controlled the country.

However, the people rebelled. They started a political movement, demanding political and economic reforms, which would later be called the Progressive Movement. Reform didn't come easy, but at the urging of President Teddy Roosevelt, Congress passed a number of new laws designed to help enforce antitrust laws already on the books, such as the Sherman Antitrust Act of 1890. The new laws included the Elkins Act of 1903, aimed at the discriminatory trade practice of railroads, and the Hepburn Act of 1906, which strengthened the Interstate Commerce Commission. During Roosevelt's two administrations, the Justice Department brought more than forty antitrust actions against the corporate trusts and won several important judgments. One judgment resulted in the break-up of Rockefeller's Standard Oil Company Trust.

The Progressive Era in American politics continued through the Woodrow Wilson administration. In politics, Civil Service replaced political patronage, crippling the powerful "political machines." Primary elections were instituted to select candidates for offices, replacing corporate deals in smoke-filled rooms. And direct elections were mandated for election of U.S. Senators, replacing selection of senators by state legislators in some states. Labor unions were legalized as a countervailing power to large corporate employers. And women were given the right to vote, creating a new political force with a different public policy agenda.

While impressive at the time, the effectiveness of reforms instituted during the Progressive Movement has been whittled away over time, par-

ticularly during the last two decades. Today, the effective concentration
of corporate power is far greater, and consequently markets are far less
competitive, than in the days of Teddy Roosevelt. In addition, today's corporations are multinational – exceeding the span of control of any single
nation, and often exceeding the size of most national economies.
Widespread corporate alliances and joint ventures add still further to the
span of control of the corporate giants. Yet today's efforts at antitrust
enforcement range from feeble to non-existent.

In spite of the "busting of old trusts," the early 1900s brought but a
pause in the industrialization of America. Henry Ford, the creator of the
Model T, is credited with opening the "modern era" of American industrialization. Invention of the internal combustion engine and discovery of
abundant supplies of petroleum fuel paved the way for "Fordian-style"
industrialization. In 1913, Ford opened his first automobile plant – complete with an assembly line. Ford didn't invent the automobile and he was
not the first to use interchangeable parts or assembly line production,
but he was the first to combine successfully the concepts of specialization and standardization in developing a new and different industry.

Each person on Ford's assembly line, as on all assembly lines since,
carried out a specific set of activities in a specific manner – performing
as little more than sophisticated machines. All parts on the Model Ts
were standardized and interchangeable. Each Model T was identical to all
other Model Ts, allowing them to be assembled piece-by-piece, step-by-
step, without regard to their individual identity. In addition, a corporate
organizational structure allowed the Ford Motor Company, formed with
capital from eleven investors in the Detroit area, to expand and eventually grow into one of the largest business organizations in the world.

Ford workers at first rebelled at the boredom of specialized, standardized, assembly line work. People were unaccustomed to such routine, repetitive, mechanistic work, and worker turnover, at 40-60 percent
per month, was intolerably high. Ford is quoted as saying that his biggest
problem was that he had to hire whole workers while all he really needed was their two hands. But in response to the worker revolt, Ford doubled his worker's wages from $2.50 to $5.00 per day. Worker turnover
dropped, productivity increased, and profits doubled within two years.
Ford discovered that the American worker could be bribed to tolerate
dehumanization. The modern era of industrialization was under way.

During the 20th century, the "Fordian" model of industrialization
spread far beyond manufacturing and today dominates nearly every
aspect of life in the United States, as it does in all so called, "developed,"
nations of the world. In the U.S., even government has become an "industry," its organization and function reflecting the same industrial, competitive mentality found in private-sector industrial corporations. The three

branches of government have evolved into complex collections of
departments, committees, commissions, and bureaus, each of which performs highly specialized functions. Laws and regulations emerge as the
final products of a long and complex legislative assembly line process.
The oft-stated goal of the government bureaus implementing our laws is
to make government "run more like a business."

The congressional committees, where all legislation begins, each
deal only with specific types of legislative issues. Lobbyists representing
special interest groups become involved in the process, attempting to
persuade committee members to support their specific interest.
Eventually, bills move out of committee to be debated by each chamber
of Congress, moving step-by-step, until eventually they are either rejected or approved by both houses of Congress and signed into law by the
Chief Executive. Making a law today is a lot like making a car. There is little opportunity for individual initiative or creativity. Each person must
function in lock-step with the legislative process.

Once a bill becomes law, the executive branch takes over. Each law
is referred to a specific government department, bureau, or agency for
implementation. Again, an assembly line process comes into play. Each
government agency performs a specific set of specialized functions in a
vertical chain of command reaching from the ordinary citizen to the
highest levels of government. There is no place for individual judgment
or creativity in addressing the questions or problems of constituents. A
veritable quagmire of rules and regulations ensure that the government
bureaucracy, not even the individual bureaucrat, is in control.
Administering a government program is a lot like running a corporation,
except administrators worry about appropriations and budgets rather
than profits or losses.

Medical care and other public services also have become increasingly industrialized and corporatized over the years. Most doctors are now
specialists, not just in surgery or internal medicine, for example, but also
in specific kinds of surgery or medicine. Many treat only patients of a
specific age or patients with a specific kind of physical ailment or mental disorder. Those who are not specialists actually specialize in being
general practitioners, in referring patients to the appropriate specialist.
Patients are passed through a medical assembly line as they go from doctor to doctor, from test to test, and from room to room in the hospital.
With specialists working on each part of the patient, the person as a
whole is supposed to be made well. We seem to think we can manufacture wellness as if we were manufacturing an automobile.

This same philosophy has come to dominate almost every type of
public service. From national defense, to public transportation, to public
schools to volunteer fire departments, each public service is provided by

a collection of specialized individuals, following standardized operating procedures, under the control of some type of centralized "chain of command." Even the larger churches have ministers, assistant ministers, music directors, and youth directors in addition to specialized staff and a host of lay committees, all working on an "assembly line for salvation." Most individuals in such organizations have very little individual choice in what they do, say, or think. Most public servants are but very small cogs in big bureaucratic machines.

Even our universities, supposed institutions of higher learning, have become little more than information factories. Most professors are no longer true doctors of philosophy, or searchers of basic truth, but have become information specialists. Universities are divided into departments by discipline, and disciplines are subdivided further into still more narrow areas of specialization. Scientists often specialize further by studying only specific types of questions within their area of specialization. They work on their own little piece of the world under the assumption that a better understanding of their piece somehow contributes to a better understanding of the whole. It doesn't – not necessarily, anyway.

Scientists analyze things; they take things apart, piece by piece. They attempt to understand the whole by breaking it up into manageable parts – like disassembling a hog carcass in a packinghouse. We gain very limited knowledge or understanding of the "whole hog" by examining only the bacon, ham, and sausage on our dinner plates. Scientists, likewise, gain very little knowledge of the whole of things by analyzing the little pieces of reality. Knowledge results from the integration of information into "meaningful" concepts, theories, opinions, notions, beliefs, or philosophies.

Knowledge requires more than observation and analysis; it requires common sense. We need not reject science as a means of gaining knowledge, but we must reject the proposition that a specific scientific process results in true knowledge. Many scientists today blindly accept often-outdated propositions and assumptions as the foundation for so called, "good science." They are relying on an industrial scientific process in an attempt to create knowledge, without examining the first principles upon which the integrity of process ultimately depends. Knowledge cannot be created in the process of assembling pieces of information. In making an automobile, the nature of the whole must be understood before the process begins. The search for true knowledge must start with a common sense understanding of the first principles that define the essence of the whole under study.

With a few notable exceptions, higher education has become little more than an industrial process of information transfer. Most students enroll in highly standardized courses of study, which specify the courses to be taken and their sequence, with relatively few, mostly restricted, "electives." Each required course in a curriculum must cover the same basic material, regardless of who teaches it, because each course builds on information conveyed in the previous course. The student moves from one stage to the next on this academic assembly line, until they emerge as a college graduate. Graduate degrees, as they say, require "more of the same."

Students emerge from this process with lots of information but with very limited knowledge – unless they have gained knowledge through their own common sense and individual initiatives. There are, however, some notable exceptions, where faculty members have rebelled against the manufacturing of education and provide knowledge experiences for their students. But, the notable exceptions only tend to validate the rule.

Few people even realize that alternative organizational models exist. It just seems natural to specialize in function, standardize procedures, and centralize decisions to make any sort of organization function more efficiently. When most people say, "let's get organized," they mean let's determine how various necessary functions fit together in some orderly sequence, let's give each person responsibility for specific processes, and let's decide who is going to be in charge. In other words, "let's get organized" means "let's form an industrial organization."

There is nothing fundamentally wrong with specialization, standardization, and consolidation of control, at least not within limits. If unchecked, however, industrialization eventually creates direct conflicts with the essential characteristics of all healthy, living organizations, including people, economies, and societies. Living organisms and organizations are inherently holistic, diverse, and interdependent. We are whole people, not a collection of parts; healthy societies are made up of different people, not copies of the same person; and healthy economic relationships must be mutually beneficial – not domineering and exploitative. Specialization ignores the value of holism, standardization ignores the value of diversity, and hierarchal control ignores the value of interdependence in healthy, living organizations.

Industrial organizations achieve their greater efficiency and productivity by removing anything that constrains their productivity and economic efficiency. Concerns of society and the environment add nothing to the economic bottom line, and thus represent constraints to industrialization that must be removed to maximize profits and growth. On the other hand, healthy living organisms and organizations have inherent controls, self-restraints, which limit their rates of growth and reproduction and define their healthy mature size or scope. These internal controls cause living organisms to devote energy to normal growth and productivity, while also devoting sufficient energy for renewal, reproduction, and regeneration to ensure long run sustainability. Living systems are able to sustain both their productivity and their regenerative capacity.

While working with Wilson & Co. in the field of advertising and merchandising, I had an opportunity to work with a genuine giant. His name was Henry Hite. Henry was one of the gimmicks we used to lure people to supermarkets to buy our bacon and hams. Henry billed himself as being eight feet, two inches tall, although the *Guinness Book of World Records* lists him at seven feet, nine-and-a-half inches. He admitted to me that he wasn't actually eight-foot-two, but said he was at least two inches taller than some other fellow who claimed to be eight-foot-even. Regardless, Henry Hite was a tall man – a genuine giant.

Perhaps the thing that distinguished Henry most among his peers was that he lived to be more than sixty years old. Most giants die young. Few survive their thirties, but Henry was lucky. All of his abnormal growth came during his teenage years; by age nineteen, he had stopped growing. Most giants keep right on growing, until their body becomes so large their vital organs can no longer support their physical bulk and they die. For most giants, the normal biological processes that would have naturally limited the size of their body fail to function. They continue to grow beyond their normal size, their health begins to decline, and eventually, they die. Henry quit growing before he grew too big to live. He pushed the limits of size, but survived.

Industrial organizations tend toward giantism. Lacking natural inherent controls, they grow uncontrollably, like a cancerous tumor, until they deplete their energy supply and destroy the life of their hosts. But the productivity and regenerative capacity of living organizations – communities, societies, and economies – depend upon an appropriate balance between specialization and holism, between standardization and diversity, and between consolidation of control and empowerment of individuals.

The myopic mentality of industrialism is at the root of nearly every social and ecological ill confronting global society today. Systematically eliminating social and political constraints, it continues an unfettered exploitation of people and depletion of non-renewable natural resources. We can not expect to sustain human life on earth for very many more decades, unless we confront the organizational villain of industrialization.

My career as an industrialist was relatively brief but very enlightening. My tenure with Wilson & Co. spanned just over three years, but by most measures of professional achievement, it could have been considered a success. I had started out in Kansas City in the merchandizing department – "games and gimmicks," I would call it later. Six months of that time was spent "on the road," from Phoenix to Boston, with Wilson's Six Horse Hitch – a team of Clydesdale horses that we took from store to store to promote Wilson meat products. I also worked with a boxing kangaroo and a singing pig. I learned a lot about the worlds of advertising and sales promotion.

I was given the opportunity to become merchandizing manager for the new branch in Atlanta, which in turn opened the door for me to become manager for a larger branch office in Detroit, Michigan. I learned a lot about the real world of corporate bureaucracy.

Most of my supervisors seemed to think I could have a bright future with the company, but I never felt comfortable with the work. Frankly, I didn't think some of the things I was expected to do were completely "honest," although they were hopefully legal. I wanted to become a person with a "good reputation," as my dad had suggested. I also wanted to make a positive contribution to society. And, I was never comfortable working in the corporate bureaucracy. Perhaps most important, however, I couldn't see anyone higher up the corporate ladder who epitomized what I wanted to become. In fact, I saw very few that I even respected.

I thought seriously about becoming a high school teacher, but soon learned that the required specialized courses in education required essentially another undergraduate degree. In the summer of 1965, I contacted the Department of Agricultural Economics at the University of Missouri about the possibility of returning to do graduate work. I was offered an assistantship to begin work on an MS degree. Deciding that "more of the same" was preferable to starting over, I accepted the assistantship, and returned to Missouri.

Maybe I didn't want to get rich after all. If I really wanted to be successful in the corporate world, I had concluded, "the company" had to take priority over family and friends, even over ethics and morality. I was not willing to sell my soul to the company. There had to be a better way. So, I went back to school. I had confronted the organizational villain and I had rejected it, before it rejected me. Although I didn't realize it at the time, the villain of industrial thinking still had a firm grasp on my mind.

Endnotes

[1] *Microsoft Encarta Encyclopedia*, 2003, "Guilds" (Redmond, WA: Microsoft Corp., 1993-2003).

[2] *Microsoft Encarta Encyclopedia*, 2003, "Industrial Revolution."

[3] Adam Smith, *Wealth of Nations* (Amherst, MA: Prometheus Books, Great Mind Series, 1991, original copyright, 1776).

[4] Smith, *Wealth of Nations*, 11.

[5] Smith, *Wealth of Nations*, 7.

[6] Smith, *Wealth of Nations*, 199.

[7] Upton Sinclair, *The Jungle* (New York: Bantam Books, 1981, original copyright, 1906).

4

FALL OF CAPITALISM:
THE ECONOMIC VILLAIN

After three years in the corporate world, my indoctrination into the religion of "laissez faire" economics was complete. I was a political and economic conservative. Barry Goldwater had been my candidate in the last presidential election. I believed in equal justice for all, but I didn't believe we could "legislate conscience." We needed a government to maintain an army, build roads, and to ensure everyone basic education, but that was about it. Government works best when it works least – when it leaves most things to the markets – so I thought at the time. As far as I was concerned, the markets worked.

"It's all a simple matter of economics," so I thought. It doesn't have to make common sense, if it makes economic sense. If it's profitable, someone will do it, even if it might best be left undone. But if there is no profit in it, no one will do it, even if it needs to be done. "That's just the way things work." To me and to other economists, this somehow made sense.

When I returned to graduate school in 1965, my economic philosophy was more conservative and free market than that of most faculty members at the University of Missouri at that time. The Department of Agricultural Economics had more advocates of the more liberal school of economics of Harvard University than the conservative school of economics of the University of Chicago. I was typically the one arguing for less government interference and more market freedom. Thankfully, I had a few good professors who were willing and able to challenge my thinking. They argued that the markets don't always work and sometimes the government has to intervene on behalf of society. They argued that unrestrained corporate consolidation would destroy the competitiveness of markets, and that industries with a few large firms would not be economically competitive and would not function in the best interest of society. Although I didn't agree with them at the time, their lessons became very helpful later in life, when I eventually began to question the

39

religion of "laissez faire" economics. Another subject that I did not fully appreciate at the time was the importance of economic history.

The roots of most economic thinking today can still be traced back to Adam Smith. The full title of Smith's classic book is *An Inquiry into the Nature and Sources of the Wealth of Nations*.[1] Apparently, it was not Smith's intention to write a book outlining a theory of economics that would remain true for all times. He was only attempting to draw some conclusions regarding the factors contributing to the creation of wealth by some nations in the late 1700s, by inquiring into the nature of the world of that time. He undoubtedly hoped that his observations would be of use for some reasonable period of time. But it was the 20th Century economists, not Smith, who turned Smith's observations into the academic discipline of economics and eventually turned economics into a religion.

Smith and his fellow "classical economists" understood that an economy is but a part of society and that a strong economy must rest upon a solid social and moral foundation.[2] They understood economics as a social science, a living science, fundamentally different from the physical sciences, such as chemistry and physics. Around the turn of the nineteenth century, however, the trend toward a more mechanistic approach to economic thinking began among those who are now called "neoclassical economists," including Walras,[3] Pareto,[4] and Marshall.[5] They wanted to add precision and quantitative rigor to economics – to make it more like the physical sciences. They developed specific sets of assumptions that allowed them to deduce cause and effect relationships and to predict the consequences of economic choices. They began to use charts, graphs, and mathematical equations to define economic relationships. The more precise logic of mathematics and statistics could then be used to derive conclusions and to either prove or disprove economic propositions. Over the years, the social, living science of classical economics was reshaped into the mathematical, statistical science of neoclassical economics.

Among the most important of the neoclassical economic assumptions are those defining the conditions under which Adam Smith's invisible hand might be expected to transform individual self-interests into the common good.[6] Perhaps the most important assumption is that of consumer sovereignty – "the consumer is king." In economics, consumers' tastes and preferences are taken as "given" – as intrinsic wants and needs, untainted by persuasive outside influences. The fundamental purpose of all economic activity is to allocate scarce resources in such a way as to maximize the satisfaction of those wants and needs. Economics makes no judgment concerning whether consumers' preferences are good or bad, or right or wrong. The job of economics is to give consumers as much as possible of whatever they want, accepting their preferences as untainted, as if decreed by God. In economics, the consumer is sovereign.

A second critical economic assumption is that markets are truly competitive. The economic meaning of competitive markets goes far beyond the common assumption that competition exists as long as two or more people are trying to buy or sell the same basic good or service. For markets to be competitive, in the economic sense, there must be "many buyers and sellers," so many that no single buyer or seller can have a noticeable effect on either the total quantity sold or price in the marketplace. Economic competition is necessary to ensure that markets quickly and efficiently reflect changes in consumer preferences back to the production level and reflect changes in costs of production up to the consumer level. Without effective competition, producing and marketing firms have the power to retain excessive profits for themselves, and thus to distort market price signals and misallocate scarce resources.

In economically competitive markets, producers must have "freedom of entry and exit" – it must be easy to get into and out of the business. If prices rise to profitable levels, new suppliers must be able to enter the market quickly, to increase quantities supplied and bring prices back down to levels of marginal production costs. If prices fall to unprofitable levels, current suppliers must be able to leave the market quickly, to reduce quantities supplied, and bring prices back up to the levels of marginal cost of production. Market entry and exit – getting in and out of particular businesses – are the means by which productive inputs and resources are reallocated to meet changes in consumer preferences. Without freedom of entry and exit, excessive profits or losses can persist, and markets will not respond to the changing needs and wants of consumers.

Economically competitive markets also require accurate market information. Consumers must be able to anticipate, at time of purchase, the benefits they are going to derive later from a particular good or service. If the ultimate value of something is either less or more than anticipated at time of purchase, the price of the good or service will not reflect its actual value to consumers. In the absence of accurate information, people end up buying things they don't need and avoid buying things that would benefit them more. Market prices then send the wrong signals to producers. Some things are produced that consumers don't want and other things consumers would actually prefer are not produced. The invisible hand of market economics must be guided by accurate information.

When all of these conditions are met, markets are said to be "perfectly competitive." The conditions necessary for perfect competition were typical of the world in the late 1700s, at least among the more "wealthy nations" targeted by Smith's inquiry. As consumers, people of those times were sovereigns. Consumer tastes and preferences could be accepted as given. There may have been strong social pressures to make people better citizens, better husbands and wives, or even to save their souls, but their material needs and wants were pretty much accepted without a lot of promoting or persuading. There was little, if any, advertising designed specifically to create wants and needs that did not previously exist.

The economy was made up of many individual buyers and sellers. Most businesses were small private enterprises, operated by a single individual or a family, with possibly a few hired workers. And one enterprise could double its production or go out of business with little impact on the overall market. Corporations were rare in those days, mainly restricted to government granted monopolies, such as the Hudson Bay Company in America, which were chartered by governments to exploit the resources of the colonies. European craft and merchants guilds were still around, but Smith criticized them soundly because of their undesirable influence on competition. Most guilds were abolished by the early 1800s.

In Smith's day, it was easy to get into or out of a business. The enterprises were small, so the investments were small. It was easy to acquire enough money to get into a business that looked promising and not too painful to liquidate one that didn't. Most buildings and equipment in those days could be put to a number of different uses, so no one was locked into doing just one thing. Since most processing and manufacturing activities were pretty basic, there were few patents or copyrights to prevent new competition for existing producers.

Information concerning the usefulness and reliability of goods and services was typically pretty basic but generally accurate. Most products were pretty basic and most transactions took place face-to-face between buyers and sellers who knew each other and did business on a regular basis. If buyers wanted to know something about a product, they could ask the producer, face-to-face. If a product didn't live up to expectations, they could take it back, and the seller would have to make it right. Misleading product promotion and advertising was largely limited to a few traveling hucksters peddling patent medicines.

Smith's observations concerning the nature and sources of wealth among nations of the time probably were quite accurate, considering the 18th century world into which he made his inquiry. Economists of the 20th century took Smith's observations from one point in time, defined the conditions under which Smith's conclusions would be valid regardless of time, and derived the basic assumptions of neoclassical economic theory. This process of developing economic theory was perfectly logical and reasonable. The problem is that none of the basic assumptions derived from Smith's observations of the economy of the late 1700s is reflected in the economy of the 2000s. The world has changed dramatically in the past 250 years, but the basic assumptions of economics have not.

Today's markets are dominated by a handful of giant corporations, not "many" independent buyers and sellers. The actions of any one of these large firms can have a major influence on overall supplies and prices in the market place. These firms have the power to set prices and to increase or restrict market supplies to protect their profit margins. Cost savings are not necessarily passed on to consumers, but can be retained as corporate profits. Consumers get the variety and quantities of goods and services that maximize corporate profits, not necessarily the variety and quantities most consumers need or would actually prefer.

These corporate firms are impacted by, and thus are keenly aware of, each other's strategies and actions. They jostle for market position, they conspire, they "play strategy games," there are not so many that any can afford to ignore the actions of any other. When the number of dominant firms in an industry drops to three or four they don't need to communicate directly in order to conspire in ways that enhance their collective profits. Such markets are not competitive.

It is not easy to get into or out of today's type of business. Total capital assets of industrial corporations today are measured in millions and billions of dollars. Most have thousands or hundreds of thousands of stockholders. It isn't easy to achieve a size that justifies a public stock offering, and without "going public," most new enterprises cannot get large enough to compete for a viable market share. Management decisions have been largely separated from ownership and the managers are not about to get out of a business for any reason short of bankruptcy. Thus, scarce resources are not efficiently reallocated among competing uses so as to maximize consumer satisfaction, as is necessary for economic efficiency.

Corporations buy and sell other corporations, they merge with and spin off other corporations, but all of this has little to do with responding to changing consumers' wants and needs. Companies may claim they are responding to consumer demands by introducing new products or "new and improved" versions of current products. But there is nothing to ensure that resources are being allocated to meet consumers' needs as long as it's difficult, if not impossible, to get into or out of business.

It is not easy to get accurate product information. Few transactions today take place face-to-face between producer and consumer. Most products go through at least a half-dozen different stages or levels of production, each represented by a different specialized organization in vertical supply chains linking production to consumption. Government regulations establish grades and standards, require specific labeling, and attempt to discourage fraud and deceptive practices. Such regulations are all reactions to the fact that consumers no longer have any real knowledge of when, where, how, and by whom their products were produced. All of this information is helpful, but none of it can possibly take the place of direct contact with producers.

Advertising has probably done more to destroy capitalism than has any single thing, perhaps other than the corporate form of organization. Advertising makes a mockery of concepts of consumer sovereignty and perfect information. Advertising today is not designed to provide information, it is designed to persuade the consumer to buy, regardless of whether they have a real need for or even want, what is being offered for sale. When the first corporate advertising agency hired their first Ph.D. psychologist to produce ads – sometime in the 1940s – the business of advertising shifted from informing to creating wants and "needs." Advertising is used to create the illusion that one product is better than another product, even when no tangible difference exists. It is designed to confuse; it is disinformation by design and a purposeful attempt to escape the discipline of market competition. There is nothing in economic theory that deals with the use of potentially productive resources – labor, money, and management – to warp and bend consumers' minds to conform to the desires of producers.

We no longer have a capitalistic economy – at least not capitalistic in the sense assumed in neoclassical economic theory. Today's large corporate organizations may have achieved the economies of scale necessary to produce lots of cheap stuff, but there is no assurance they are producing the right stuff to meet the real wants and needs of consumers. The microeconomic theories of today are hopelessly and dangerously out of date. The invisible hand of Adam Smith's competitive capitalism is no longer relevant to today's "free market" economy.

Many people seem to believe that the government is looking out for the economic good of the public. Corporations are always complaining about excessive government regulations. However, the government does very little to ensure that economic needs of the society as a whole are met. "Macroeconomic theory" deals with economic issues at the aggregate or national level. However, one of the wisest of my economics professors, Harold Breimyer, was fond of saying that the concept of macroeconomic theory was a myth, because there are no economic theories relating to the economy as a whole.

The "macro-economy" is nothing more than the sum of all of the individual "micro-economies." In economics, the performance of the economy overall is simply the sum result of decisions made by individual consumers and producers. In economics, the national economy has no characteristics other than those embodied in the separate economic entities, the individuals and organizations, of which it is composed. The economy as a whole is nothing more than the sum of its parts. In reality, the purpose of macroeconomic policies is to promote the continuing growth of private economic enterprises.

A good part of the first college level course in economics that many people reading this book didn't take and most others didn't understand dealt with macroeconomic policies. Macroeconomics can be even more abstract and boring than microeconomics, particularly to a young college student who has no direct contact with government, except perhaps filing a tax return. More people probably have a reasonable understanding of how markets work than of how government policies affect markets. Lacking this understanding, people can be led to believe that if the economy is growing, it must be healthy, and the government has everything under control. Or they may believe that if they have just been laid off, their company has just dropped its pension program, or their job has been exported to another country, there must be a problem with government economic policy. Many people actually believe that the Chairman of the Federal Reserve Board is taking care of their economic future.

Most people just don't understand enough about economics to understand or explain what the government is doing, right or wrong, with respect to economic policy. Those of us who have had courses in economics probably didn't really understand half of what was being taught at the time, even if we had a good instructor. As with microeconomics, knowledge of basic macroeconomics can be powerful defense against the simple dogma regarding the role of government. I will attempt to make the following mini-lesson in macroeconomics a bit more interesting by mixing it in with a bit of American economic history.

Macroeconomic policies are divided into monetary and fiscal policies. Monetary policy relates to providing adequate currency to facilitate market transactions and to finance the investments needed for continuing economic growth. Sometimes the money supply is expanded by more than enough to finance current investment demand, in hopes that "easy money" will stimulate new investment. However, increases in the money supply that are not followed by expansion in economic output only result in inflation in overall price levels, which only serves to diminish the value of current investments. Using monetary policy alone to promote economic growth has proven to be a risky strategy.

Prior to the Great Depression of the 1930s, the only significant economic role of government was to manage the supply of money. The functions of money, which includes all forms of currency, are to serve as a medium of exchange, as in buying and selling, and to make it possible to save or store economic value. The government is the logical entity to "coin" or create the currency needed for exchange and savings.

Interest is the "price of money." Interest is a reward to those who choose to save and a cost to those who choose to spend or invest. Interest rates provide incentives to lenders, ration loans among borrowers, and thus, balance savings with investment – just as other prices balance supply with demand. Inflation in overall price levels, after which the same things then cost more, can reduce the exchange value of money over time, and thus, can reduce the "net interest rate," or price of money. So inflation must be kept under control, or at least at a predictable rate, if interest is to serve as an effective price of money. The government is the logical institution to maintain the exchange value of money, by keeping inflation in check, ensuring that interest rates reflect a reasonable price for saving and investing money.

Prior to the Great Depression, there was no role for the government in "managing the economy." Economists assumed that a free market economy would always be self-correcting. The economy was expected to go through cycles of expansion and contraction but would always return to some long run equilibrium or stable trend. When businesses became unduly optimistic, economic reality would bring forth a recession. When businesses became overly pessimistic, the economy would respond with an automatic recovery. Lower prices and profit would clear the markets of surpluses caused by overly optimistic production, and higher prices and profits would spur added production during periods of scarcity induced by undue pessimism.

However, the depression of the '30s shook the foundations of macroeconomic policy. Most of the money in circulation then, as is still true, was neither coins nor paper currency, but was money in the form of bank deposits. Banks are allowed to "create money" by making loans while holding only a fraction of the total amount of their loans in the form of customer deposits. For example, a bank with a million dollars worth of customer deposits might have five million dollars in outstanding loans. Borrowers are free to spend the full amount of money lent to them by the bank, so the bank "creates" four additional dollars for each dollar they have on deposit. As long as those with deposits in the bank don't all ask for their money at the same time, the bank will have plenty of money on hand to meet their needs.

As the economy slowed in the 1920s, workers lost their jobs and began to draw on their deposits in banks to cover their living expenses. Banks had loaned all the law would allow and thus had to call in loans to maintain the required loan to deposit ratios. Since a bank might have loaned out five dollars for each dollar of deposits, a thousand dollars drawn out of depositors' savings account could have forced a bank to reduce its outstanding loans by five thousand dollars.

Banks that "create" money when they receive new deposits must "destroy" a like amount of money when those deposits are withdrawn. The government failed to provide enough additional money to banks to offset the sharp decline in money triggered by customer withdrawals. In fact, the federal government chose this time to reduce government spending in order to balance its budget, which took still more money out of circulation.

Many banks couldn't call in loans fast enough to meet the withdrawal demands of depositors and were forced to close their doors. There was no Federal Deposit Insurance Corporation (FDIC) back then, so depositors simply lost whatever money they had in the bank. A few bank failures triggered still more withdrawals, as customers scrambled to withdraw their deposits before their bank failed. As the depression deepened, an increasing number of borrowers were forced to default on their loans, leaving banks with no means of recovering enough money to pay off their depositors. Panic set in, causing runs on banks nationwide. Bank failures were but one symptom of the depression that engulfed the nation for more than a decade, but they most clearly symbolized a failed economic policy of the government.

The Great Depression caused economists to rethink the whole concept of macroeconomics. The government had failed to maintain an adequate money supply and thus had not been able to stabilize the value of money. The economy did not correct itself automatically. In the depths of the depression, there was virtually no demand for money to invest at any interest rate. With no new investment, there was no new employment, and with no new employment, there was no new income to spend, and thus, no reason to invest. The economy didn't bounce back. It was caught in a downward spiral from which it was incapable of recovering on its own.

In the mid-1930s, a British Economist, John Maynard Keynes, published another landmark book on economics, *The General Theory of Employment, Interest, and Money*.[7] In it, he outlined his theories relating to why economies were not necessarily self-correcting and suggested government policies that could be used to bring about the necessary corrections. His basic conclusion was that monetary policy alone could not ensure recovery from deep recessions, so governments would have to take the lead by spending money to stimulate economic recovery.

Keynes' theories supported policies that had already been initiated in U.S. and Great Britain. Franklin Roosevelt's New Deal programs included a number of government spending projects designed to kick-start the economy by providing government jobs. The government spent more money than it collected in taxes and created new money to make up the difference. Government jobs would create new income for workers, new income would create new demand for goods and services, the new demand for goods and services would trigger new private investment, and the downward spiral would be broken.

It eventually took the massive deficit spending by governments in financing World War II to pull the U.S. and world economies out of the depression. However, Keynesian Economics was validated, and it survives today as the foundation for contemporary macroeconomic theory. Management of taxing and spending by the federal government is called "fiscal policy." The federal government can either run a national budget deficit, meaning that it spends more than it takes in, or run a surplus, meaning that it takes in more than it spends. Of course, it could also balance the budget, but that rarely happens.

The President and Congress make fiscal policy – to the extent that fiscal policy actually is made rather than "just happens." Typically, federal budget deficits are more a result of lack of political will to balance the federal budget than of any intentional fiscal policy, and federal budget surpluses are far more likely to be accidental than planned. Nonetheless, budget deficits tend to stimulate the national economy by creating new demand for public goods and services, and budget surpluses dampens the economy by reducing the demand for public goods and services.

Many people seem to think that the federal government should simply balance its budget, as families and state governments are expected to do. It's difficult to defend federal fiscal policy as it has been practiced, or not practiced, for the past several decades. However, in times of deep recession, the government has no means other than deficit spending to stimulate economic recovery. The federal government is responsible for maintaining stability of the national economy; individuals and state governments are not.

Monetary and fiscal policy must work together to ensure effectiveness of either. It makes no sense to maintain high interest rates and large government deficits, as during the Reagan years. Likewise, it would make no sense to have large government surpluses while interest rates are cut to stimulate an economic recovery. Pursuing conflicting monetary and fiscal policies is like pushing on the brake and accelerator of a car at the same time – it's pretty hard on the economy.

However, neither monetary nor fiscal policy deals with the interests of society as a whole or the interests of the nation, in any sense other than as a collection of individual economic enterprises. Nothing in macroeconomic policy protects the social fabric of our society from the stresses and strains of over-vigorous pursuit of our economic self-interests. Nothing in macroeconomic policy protects either natural or human resources from economic exploitation. Macroeconomic policies simply facilitate the pursuit of our individual, short-run, self-interests.

Macroeconomic policy seeks to ensure that the economy continues to grow at its maximum feasible rate. If our approach to economic development exploits the natural environment, as does current industrial economic development, the effect of "good" macroeconomic policy is maximum exploitation. If our approach to economic development is inequitable and unjust, as is industrial economic development, the effect of "good" macroeconomic policy is to maximize the inequity and injustice. Macroeconomic policies do nothing to ensure social equity or justice or to protect the natural environment. Our national economic policies may not be the root cause of exploitation, but they facilitate the process of exploitation.

After receiving my Ph.D. at the University of Missouri, I took a position as an Assistant Professor of Economics at North Carolina State University. The NCSU faculty at that time was dominated by the Milton Friedman, Chicago School of economic thinking – very conservative, "laissez faire," and free trade. I learned a lot of neoclassical economic theory in both formal and coffee room discussions with other faculty members during my six-plus years in North Carolina.

I also learned a lot from my experience as a classroom teacher. I accepted responsibility for teaching the junior-senior level course in Agricultural Marketing at NCSU, which I renamed the Economics of Agricultural Markets. I didn't want to teach marketing strategies; I wanted my students to understand how agricultural markets work so they could develop their own strategies to fit their particular needs. There is no better way to truly understand a subject than to teach it. Although my Ph.D. work was in the area of marketing, I gained new understanding and insights every time I taught the course. Each year, the course became more rigorous and each year my students evaluated the course more highly. Good teaching requires a high level of understanding.

Most economists seem to believe that the "free market works." They admit that we no longer have "perfect competition," but argue instead that we have "workable competition." They have convinced themselves that the necessary conditions of economic competition really aren't necessary. Perhaps there aren't as many buyers and sellers as one might like, but there are enough to ensure competition, they say. After all, business firms have to be large to achieve economies of scale. If we limit the size of firms, their costs would be higher, and prices to consumers would be higher as well. Perhaps market information isn't perfect, but it costs money to provide information, so consumers are better off with less information and lower costs. It may not be as easy to get into or out of business as we might prefer from a competitive standpoint, but again we have to consider the tradeoffs. Large investments are necessary because firms have to be big to keep costs down. Patents, copyrights, and licensing are necessary because research and development is costly. Businesses must be granted a monopoly on new discoveries or they won't make adequate investments in research and development – so the arguments go. In the world of the free market economist, "imperfect competition" is almost as good as "perfect competition," and "the markets work" – as a matter of faith.

From North Carolina, I went to Oklahoma State University. OSU was only slightly less conservative than NCSU. At the time, I had no real incentive to question the prevailing economic wisdom of free markets. After a couple of years, however, a few economists with liberal leanings began to infiltrate the OSU faculty, significantly increasing the interest of coffee room discussions. One day while checking out a sidewalk bin of discounted books, I came across a copy of *The Communist Manifesto* by Karl Marx and Friedrich Engles.[8] I had a Ph.D. in Economics but I had never read anything of significance by Karl Marx. Buying such a book in Oklahoma was a bit like buying pornography; I felt like I should carry it to my office in a brown paper bag. But I wanted to see what Marx had actually written about communism.

As I began to read Marx's description of the ultimate consequences of capitalism, I began to see the ills of modern American society unfolding in my mind. To my surprise, Marx sang the praises of capitalism – with its specialized production systems, ownership of private property, and reliance on free markets – as the most efficient organization of the means of production that humanity had ever devised. He wrote that the material achievements of capitalism in one century had exceeded the achievements of humanity in all previous times. But he also wrote of the inevitable consequences of capitalism's insatiable need for continual innovation and growth and of its necessity to reduce all human relationships to monetary transactions.

Marx claimed that capitalism eventually would reduce the workers to little more than sophisticated machines, with little, if any, consideration given to their uniquely human qualities. He wrote of the necessity for capitalistic industry to break the ties of people within families, within communities, even within nations, so that people would respond "appropriately" to economic incentives for innovation and growth. Marx wrote of the insatiable need for industry to grow, always to produce more, to meet the needs of the present, but always to promise more for the future, thus providing an incentive for continual change and unending growth. Simply stated, capitalist firms must continually search for new resources and new workers to be exploited, in a never-ending quest for more and cheaper stuff.

If a capitalistic economy stops growing, incentives for new investment decline, employment declines, incomes decline, demand for new goods and services decline, profits decline, production declines, and investments decline still further. The economy falls into recession, and eventually sinks into depression, if nothing is done to stimulate new growth. A capitalistic economy must either grow or die. Thus everything else, including people and natural resources, must be sacrificed for the sake of economic growth.

Marx claimed that exploitation of workers eventually would create a two-class society. The Bourgeois, or capitalists, who control the means of production, and the Proletarians, or working class, who must either work, for whatever wage under whatever conditions the capitalist offers, or starve. Marx claimed that the Proletarians eventually would find life intolerable and would revolt against the Bourgeois, thus opening the door for political and economic reform.

Marx's solution to all of the ills of capitalism was communism. With the collapse of the former Soviet Union, however, communism lost its credibility as a viable economic and social system. History had shown Marx's prescription to be wrong, or at least inadequate, although history has validated his diagnosis.

The failure of communism, however, does not nullify the logic of Marx's observations concerning the ultimate outcome of unbridled capitalism. I was a never a communist – I never thought that communism could work. But, I could see a lot of wisdom in what Marx had to say about capitalism. I felt then, and still feel, that any wise capitalist should heed his warnings. We must address the inherent weakness of capitalism identified by Marx, if we are to build a sustainable capitalistic economy.

In the past, our government had taken some bold steps to restrain the capitalistic corporations and had moderated their impacts both on workers and on society. In the U.S., some economic functions, such as transportation and education, had always been carried out under government oversight, if not outright ownership or control. In addition, when the economy had failed to serve society, as in the early 1900s, Americans had broken up the corporate trusts and had encouraged labor unions to countervail corporate power. They had protected workers against excessive exploitation by establishing worker health and safety standards and had provided for unemployment compensation. For old people without adequate savings who could not work, the government had established Social Security Insurance. In principle at least, neither workers nor consumers in the U.S. were to be exploited for the sake of profit and growth, a principle difficult to enforce but nonetheless important.

Perhaps even more important, Americans had established a general principle that we would redistribute income and wealth, as necessary, to preclude the development of a two-class society, thus heading off the revolution predicted by Marx. It had taken the Great Depression of the 1930s to bring home the reality of a possible economic collapse. But the government eventually had responded wisely, intervened in the economy, and thus, had preserved capitalism, at least for another generation. Without these "socialistic" influences within the American democracy, Marx's predictions quite likely would have come true.

We are told that our capitalistic economy is "what made America great." But, without the social and ethical restraints of American government and society, the economy would likely have degenerated into an oppressive system of exploitation several decades ago. Marx did not envision a nation with the moral courage and collective will to restrain the natural tendency of capitalism toward oppression and exploitation. However, in the past few decades, we have begun systematically to dismantle all ethical and social restraints on raw capitalism. We are now again moving quickly on the path toward the final stages of unbridled capitalism as envisioned by Marx – toward exploitation, inequity, chaos, revolution, and ultimately, economic collapse.

Most economists today have no comprehension of the raw economic and political power of today's giant corporations. Competition is no longer "workable" – it simply doesn't exist, at least in an economic sense. Entry of successful new firms into established industries is virtually impossible, as is the orderly exit of firms. Advertising has evolved from persuasion to brainwashing, and consumer sovereignty has been replaced by corporate sovereignty. Capitalism has been replaced by corporatism. There is no theoretical foundation for the corporatist economy in any economics textbook. Economists simply deal with the economy "as if it were" competitive capitalism. Clearly, it is no longer either competitive or capitalistic.

In Oklahoma, I worked mostly with cattlemen. My main job was to help cattlemen anticipate changes in cattle prices so they could make appropriate adjustments in their operations to maximize profits while avoiding large losses. After learning that no one, including me, could forecast cattle prices with any degree of accuracy, I taught cattlemen to use the cattle futures markets to hedge against unexpected adverse price changes, by taking risk positions in future markets to offset the risks they were taking on cattle prices. I was paid by the taxpayers, so I was always aware that taxpayers, not just farmers, were supposed to benefit from my work. If I did my job well, cattlemen would expand production to lower the peaks of rising prices and cut production to raise the valleys when prices were falling. Prices would be more stable, benefiting both consumers and producers. By hedging their prices, cattlemen could manage their production more effectively, and thus, reduce their overall costs of production and keep down the price of beef at the retail level. I was helping the markets work better, for the good of the public.

At the time, agricultural commodity markets were still pretty competitive, even in the classical economic sense of competitiveness. There were so many farmers and ranchers that no one producer of any single commodity had any measurable impact on the overall markets for the things they sold. It was still fairly easy to get into or out of farming, market information was pretty good, and consumers were still king in the food market. Markets were less competitive in the agribusiness sector, where farmers bought their inputs and sold their raw commodities, but farm level markets exhibited many of the characteristics of economic pure competition.

As an economist, I could see first hand how this kind of free-market economy benefited consumers, at least in the short run, but it sure was hard on farmers. The relentless competition of free markets forced farmers to farm more land, produce more livestock, to expand their operations year after year, just to stay in business. So much of the profit was being competed away from farming that there was too little left for farmers to care for their families or for the land. Farmers were forced to expand production to compensate for narrower profit margins. Farmers couldn't afford to care about their neighbors because sooner or later, one way or another, they would have to have their neighbors land, just to survive. They couldn't afford to pass the land on to the next generation, "as good as they found it," because they had to make it produce as much as possible at the lowest possible cost, just to survive. They were told they had only two choices: they could either get bigger or get out of farming. Those who couldn't get bigger and choose not to get out of farming were creating a new rural underclass. Perhaps for the first time, I was beginning to understand that even a competitive free market economy didn't necessarily work for the long run benefit of rural society.

I was seeing first hand what Marx had written about, capitalism left to pursue its natural course leads to a two-class society – those with capital and those without. Marx thought that capitalism ultimately would be destroyed by revolution, by a struggle between classes. What Marx, and apparently few others, had anticipated was that capitalism, left to pursue its natural course of development, ultimately would use up its natural and human resources and thus eventually would destroy itself. Unbridled capitalism quite simply is not sustainable.

Somehow, I was predisposed to see what was taking place in the economy of agriculture from a perspective different from most of my colleagues - at least during the last half of my professional career. I was willing to question and to doubt the inevitability and desirability of the capitalistic industrial process, while most others were not. I had grown up on a farm, I had operated a small business, and I had worked for a large corporation. All of these things helped shape my later thinking. I also had been able to gain some unique insights from agriculture that I am confident apply to the rest of the American economy as well.

We are currently in the midst of a great economic experiment. We have reason to believe that this experiment is leading to destruction of the natural environment, degradation of human civilization, and will eventually spread another "age of darkness" across the world. To stop this experiment, we quite likely will have to dismantle the global corporate structure, and world leaders seem unwilling even to consider such an action, at least not in response to anything short of a global catastrophe.

The people of America and the world are firmly in the grasp of an economic villain - a villain such as the world has never before confronted. We must use every means at our disposal to break free of its grasp. It's time for economic revolution, a return to common sense, before it's too late.

Endnotes

[1] Adam Smith, *An Inquiry Into The Nature and Causes of Wealth Among Nations*, ed. Adam and Charles Black; and Longman, Brown, Green & Longmans (London: J. R. McCulloch, 1828, original copyright, 1776).

[2] Henry George, "The Single Tax: What It Is and Why We Urge It," in *Christian Advocate*, 1890 (New York: Robert Schalkenbach Foundation, 1990).

[3] Leon Walras, *Elements of Pure Economics*, translated by William Jaffe (New York: Irwin Publishing Company, 1954).

[4] Vilfredo Pareto, *Mind & Society* (New York: Dover Publishing Company, 1935).

[5] Alfred Marshall, *Principles of Economics* (New York: Macmillan Publishing Company, 1920).

[6] C. E. Ferguson, *Microeconomic Theory* (Homewood, Ill.: Richard D. Irwin, Inc., 1969), 222-225.

[7] J. M. Keynes, *General Theory of Employment, Interest, and Money* (New York: Harcourt Brace, 1936).

[8] Karl Marx and Friedrich Engles, *The Communist Manifesto* (New York: Pocket Books, Simon and Schuster, (1964), original copyright, 1848).

5

RISE OF CORPORATISM:
THE SOCIETAL VILLAIN

During my academic career, I spent most of my time working with farmers on issues related to economics. I taught several different courses in agricultural economics over the years and always had significant research responsibilities, but my primary responsibility was extension education – "taking the University to the people." In the early 1980s, I moved from Oklahoma to take a position as Head of Extension Agricultural Economics at the University of Georgia. American agriculture was in crisis at that time, and I was beginning to get the feeling that we agricultural economists were at least partially responsible.

We in the agricultural universities had helped lure farmers into adopting new technologies with the promise of profits, knowing that any profits they made would be short-lived. We knew that after each new round of technology they would have to invest more money, farm more land, feed more livestock, buy more equipment, and hire more labor – just to survive. We knew that with each new round of innovation some farmers would be forced to fail, not because they were necessarily inefficient or reluctant to change, but just because some had to fail so others could "succeed." The survivors always needed more land and the only land available belonged to their neighbors. My common sense told me something was fundamentally wrong about this process.

The farm crisis was severe in Georgia. During the U.S. farm exports boom of the 1970s, many small farmers had been encouraged by government agencies to borrow money and expand their operations to take advantage of the new opportunities to feed the world. However, high interest rates during the early 1908s had inflated the value of the U.S. dollar in international markets causing agricultural exports to fall like a rock, bringing farm commodity prices down with them. Many farmers were caught with large loan commitments at high interest rates with no means of repayment. Bankruptcies and foreclosures were common and a few farmers even committed suicide when confronted with the inevitable loss of their land. In an effort to help, I was able to get a government grant for our department to carry out one-on-one financial counseling sessions with farmers. It was during these sessions with farmers that I came face-to-face with the failings of my profession. It is one thing to understand a crisis in the abstract, but it is quite another to be confronted with the reality of a crisis in the faces and the words of real people sitting just across the table from you.

During these sessions, I began to understand what was wrong in the farm economy. Only later would I understand that the same thing also was wrong with the economy in general. During the 1980s, I saw farms going broke and I saw farm families broken as they were forced off land that had been farmed by their families for generations. We met with groups of farmers who were angry with their government, with their universities, seemingly with everyone; they felt they had been betrayed. Some of the so-called farm experts blamed farmers' financial problems on poor management, which in some cases was true. But, I began to realize that the farmers with the greatest financial difficulties were those who had been doing the things that the universities, USDA, agricultural lenders, and other farm experts had been encouraging them to do; they had reason to feel betrayed. The farmers who were in the biggest trouble were the "good farmers" – the innovators and early adopters – who had borrowed heavily during the 1970s to modernize and expand their farming operations. These "good farmers" were being forced out of business.

I also saw the boarded-up storefronts in small towns as I traveled about the state; rural communities were literally withering and dying. Farmers were still producing as much or more than ever before. But, there were no farm profits to be spent in town to buy new machinery, production supplies, new cars or pickups, or new shoes and clothes for the kids. Anything that wasn't absolutely necessary would have to wait for better times, and for many, better times in farming were not going to return. Families who were forced off the farm looked first for work in town, further depressing the local labor market. But eventually they were forced to leave the local community for somewhere else – anywhere they could find a job.

The land is still farmed, but by larger, more industrialized farm operations. These larger operations buy and sell in larger quantities than do smaller family farms. They get premium prices for sales and quantity discounts for purchases when they bypass the local community and deal directly with the large agribusiness processors and input suppliers. The local equipment dealers, feed and fertilizer dealers, and marketing agencies suffer along with the family farmers. Soon, there aren't enough kids to support a local school, not enough people to justify a medical clinic, or even a doctor, and not enough families to fill the church pews. Too few are left with the time or energy to support a volunteer fire department or even to run for the town council. A community dies – or rather is killed. Somehow, the death of rural communities was justified as a simply a matter of economics.

Why should people in general be interested in the things I have seen happen to farmers and rural communities? The same things have already happened to nearly every other segment of the American economy; that's why. This is the same process by which the craftspeople of the past were replaced by factories. It's the same process by which "mom and pop" grocery stores were replaced by supermarkets, small dry goods and hardware stores were replaced by the giant discount stores, and locally owned restaurants were replaced by franchised fast food joints. This process is not about enhancing the quality of life of people – it is about producing more cheap stuff. The underlying assumption is that consumption is all that matters, and if people in total have more cheap stuff, society in general will be better off.

This process of industrialization is about production, not people. It doesn't seem to matter that people suffer from depression or even commit suicide because their businesses have failed. We label them as sick, irrational people. "Why don't they just accept the inevitable and get on with life?" It doesn't seem to matter that families are destroyed through disappointment and conflict emerging out of economic failure. It doesn't seem to matter that communities are destroyed and quality of life is diminished, even for the survivors of these "economic adjustments." It doesn't seem to matter that people abuse each other and abuse the land and other natural resources in their desperate attempts to survive the turmoil that is not of their own making. Consumer prices fall and the overall economy keeps growing. That's all that seems to matter to those who are in positions of influence and power. The process of consolidation is a fundamental characteristic of the capitalistic economic system, and America seems committed to letting capitalism run its destructive course.

Economic consolidation is the process by which the economy is brought under corporate control – by which the economy moves from capitalism to corporatism. As consolidation leads to larger and larger business organizations, it becomes harder to amass sufficient capital to achieve potential economics of scale. Thus, businesses that expand fast enough to survive are forced to incorporate in order to accumulate sufficient capital. At first, corporations tend to be family corporations, making capital accumulated during one generation available to the next generation within the same family. As the enterprises become still larger, family operations may find it difficult to accumulate sufficient capital to finance further expansion. Eventually, the successful corporation will have to "go public," selling shares to the public to raise additional investment capital. As new public corporations continue to grow, they begin to look to consolidation with other corporations as a means of expanding even faster. And as corporations grow even larger and fewer, through the process of consolidation, fewer firms will control an increasing share of total output, and markets will become less and less competitive and corporations gain more economic and political power. This is the process by which capitalism is transformed into corporatism.

It would take me a long time to fit all of the pieces together, but the direction of my professional career was forever changed by the things I began to understand during those face-to-face sessions with the farmers of Georgia. From that time on, I have been a critic, and often an opponent, of the "establishment." The establishment - meaning universities, government agencies, civic organizations, and corporations - has been promoting an economic system that ensures the continual pursuit of economic efficiency and economic growth, and thus promotes the transformation of capitalism into corporatism. I didn't start using the term "corporatism" until many years later, but little by little, I began to see the failure of neoclassical economics reflected in the rise of a corporatist society.

Webster's dictionary defines corporatism as "the organization of society into industrial and professional corporations or groups which serve as organs for political representation and which exercise control over persons and activities within their jurisdiction."[1] The essence of corporatism is people functioning as members of groups rather than as individuals. Corporate group members select leaders to represent them in fulfilling their various political, economic, or civic responsibilities. Corporate leaders, acting in behalf of the members, are able to exert a degree of influence and control over the interests of members of corporate groups. Corporatism, in the generic sense, is a natural consequence of the industrial mind-set that permeates the whole of contemporary society. As we have come to rely on specialization as the primary means of achieving greater efficiency, we have learned to function as members of various special-purpose corporate organizations.

In business, stockholders specialize by providing investment capital, while others provide labor and still others provide management. The chief executive officer of the corporation represents the stockholders in carrying out the day-to-day operation of the corporation, under the guidance of a board of directors. Workers in corporate organizations - businesses, government, or public service agencies - are members of various departments, divisions, or work units and are represented by their respective department heads, directors, or unit leaders. Workers also may choose to organize into labor unions and select union leaders to represent them in negotiations with management.

In communities, people specialize in helping others fulfill their public responsibilities by forming various nonprofit, religious, and political organizations. The rest of us join chambers of commerce, neighborhood associations, and various environmental and social justice organizations, which represent us in local civic matters. In religion, we become Catholics, Baptists, Presbyterians, Lutherans, Jews, and Muslims and are represented by priests, ministers, rabbis, or other clerics. In politics, we become Democrats, Republicans, Libertarians, or Greens and are represented by the leaders of our party. In some cases, we may be the one who represents other members of our groups rather than the one who is represented by others. Regardless, in contemporary American society, members of organizations rarely represent themselves.

Political parties, labor unions, special interest organizations, and nonprofit organizations can be just as "corporate" in concept as is any corporate business organization. When the leadership of such organizations become separated from its membership, meaning when members do not participate directly in organizational decisions, the leadership almost invariable puts perpetuation of the organization ahead of fulfillment of organizational purpose. Such an organization can be just as ruthless in its pursuit of increased membership, increased revenues, and increased salaries for organizational officials as is any private corporation in pursuit of profits. And as nonprofit organizations grow in membership, members find it increasing difficulty to participate directly in organizational decisions – the organization becomes "corporatized."

I should also hasten to point out that business organizations are not all necessarily corporate in nature, even if they are incorporated. For example, a family-held corporation, where the family members participate in management decisions, may be managed the same as a family-owned, individual proprietorship. The personal values of the family will be reflected in the decisions of the corporation. A corporation that is closely held, where specific individual stockholders can actively participate in decisions of the corporation, is no different from a partnership among individuals. The decisions will reflect the personal values of the stockholders. It's only when the stockholders become so many that individuals are unable or unwilling to impose their personal values on corporate decisions, that the organization becomes "corporate" in the generic sense of the word. When stockholders own stock for the sole purpose of increasing their wealth, as in the case with ownership by most mutual funds and pension funds, corporate managers have no choice other than to maximize profits and growth. In fact, they have a legal responsibility to do so.

Corporations are not real persons, regardless of the opinions of the Supreme Court. Corporations are not born but are created. They are not citizens, and thus, have no right to vote. They have no human rights, because they are not humans and have no right to equal protection under the law. Corporations are passionless, emotionless, inanimate legal entities, designed to pursue a narrowly defined purpose. They have no heart, no soul, no family, or community. They can have neither human concern for humans nor human ethics or morality. Anything of real human concern that interferes with the achievement of their narrow economic purpose is viewed as an obstacle that must be removed.

Some corporations attempt to integrate the human values of owners and workers into their corporate management philosophy. For example, Interface Inc., a large Georgia carpet manufacturer, and Fetzer Vineyards, a large California winemaker, openly advocate managing for long run sustainability – using economic, environmental, and social measures of success.[2] They call it managing for "the triple-bottom line."[3] The executives of such corporations should be commended for their commitment to humanity, but they are the exceptions rather than the rule. Executives of such companies publicly admit to pressures from stockholders to justify each decision in terms of economics and to give priority to profits over the environment and society. In general, as the number of stockholders becomes large, managers find it increasingly difficult to agree on giving priority to any common objectives other than profits and growth.

Of course, some individuals and family corporations are just as greedy and ruthless as any public corporation, but they have a choice to be otherwise. They deny their moral and social responsibilities because they choose to, not because they are forced to. The lack of choice is what separates most public corporations from individuals. The managers of most publicly held corporations have no choice – they must maximize profits or lose their job. The choice of corporate management as a career is a choice to manage for profits and growth.

Capitalism depends upon individual ownership, individual decisions, and individual acceptance of responsibility. Corporatism, on the other hand, is characterized by collective ownership, delegation of decisions, and separation of both decisions and ownership from responsibility. After decades of unchecked corporate consolidation, America's markets are no longer economically competitive. It's no longer a matter of markets being "imperfectly competitive" or "sufficiently competitive" – they simply are "not competitive." In America today, most of private property is owned by corporations, not individuals, most business decisions are made by corporations, not individuals, and, neither corporate owners nor managers accept full responsibility for the actions of corporations. There is no assurance today that the pursuit of individual self-interests is somehow being transformed into the best interest of society.

Another foundational principle of capitalism is minimum involvement of government in the economy. In America, however, the government and the economy have become essentially "one and the same." Corporate interests permeate every aspect of government, from the making of laws to the delivery of basic public services. It's virtually impossible to campaign successfully for any major office without corporate financial backing. The corporations have gained so much influence in government that not only does government fail to constrain corporate greed but government has become a tool of the corporate exploiters.

Corporate domination of politics is not new. Teddy Roosevelt and the Progressive Movement confronted corporate political influence in America in the early 1900s and at least slowed its growth. Dwight Eisenhower warned of returning corporate influence on the military sector in the 1950s. Many seemed to listen, but no one did much about it. Since the early Reagan years, direct corporate influence on political, social, and personal matters has been not only unchecked but also encouraged. Today, the corporatist movement seems to be moving into new territory, even beyond the excursions of a century ago.

Corporations now affect nearly every aspect of American life. Corporations decide what we can buy in the stores, what types of entertainment we are offered, what kinds of cars we can drive, and how we will communicate with each other, if we choose to participate in mainstream society. Corporations decide what kinds of schools our kids will attend, what they eat and drink there, what type of church pews we will sit in, and even decide what we think we want and "need," if we let them.

All corporate business decisions are intended to further the pervasive corporate goals of profit and growth. Increasingly, corporations are finding that their best investments are those made in shaping the values of modern society to accommodate the economic needs of the corporation, rather than simply producing things to meet the needs of society.

The potential societal gains from industrialization – specialization, standardization, and consolidation of control – were largely realized several decades ago. The corporate model no longer has any particular economic advantage in organizing for efficiency of production. In fact, as we move farther into the post-industrial era of development, with its focus on information and knowledge, the corporation is becoming increasingly obsolete. But corporations have no natural life span. They do not die a "natural death" – no matter how old or obsolete they become. They are designed to live forever and to do whatever must be done to satisfy their

never-ending need for profits and growth. Thus, corporations quite naturally have turned from resource utilization to resource exploitation as
their ability to meet the real needs of society has declined.

The corporatization of American society is perhaps more clearly
understood with respect to politics than to any other aspect of society.
One doesn't need to analyze reams of campaign contribution statistics
to know that corporations finance most political campaigns. The only
close contenders to corporate businesses in political contributions are
labor unions and political action committees (PACs), which are simply
different kinds of corporate groups, focused on shaping government to
meet the specific desires of their particular interests.

By far, the biggest costs of running a major political campaign are
associated with mass media, including television, newspapers, radio, and
mass mailings. Political consultants use tactics honed by corporate advertisers to promote and package their candidate – to cover up serious
flaws and create illusions of positive attributes. They attempt to create
false threats and unfounded fears in the minds of voters, so their candidate can promise to remove those threats and fears, if he or she is elected. They attempt to erase real threats and fears from the minds of voters,
if their candidate has no "saleable" solutions to promise the voters.

A political campaign is not about letting voters learn as much as possible about the candidate; it is about carefully packaging and promoting
the candidate, so he or she will get elected, regardless of their qualifica-
tions. Such campaigns cost money, in major elections, lots of money.
Corporations have lots of money to spend, and lots of things that they
want politicians to do for them. One of the worst and potentially most
destructive rulings ever made by the Supreme Court was when they
decided that the principle of free speech included the right of corporations, as "unnatural citizens," to contribute to political campaigns.

Politicians know there is a major problem, but they don't know how
to solve it. Periodically they make feeble attempts at campaign finance
reform, but always leave plenty of corporate loopholes. They don't see
how any national politician can openly oppose corporate influence
without being defeated in the next election. If the corporate opponents
are defeated, the newly elected corporate advocates will simply undo
anything the opponents might have done previously, and the situation
will be the same as before. Why should they sacrifice their political
career, if in the end it isn't going to do any good?

Individuals also contribute lots of money to campaigns, but it is far
more difficult to raise large sums of money from many individuals than
from a few large corporations. People know they have little hope of
influencing the political process in their favor, acting purely as individuals. That's why many join political action groups to represent their social

and ethical interests and also why many also support the "right" of the corporations for whom they work, to represent their economic interests in the political arena.

People, as individuals, have largely dropped out of the political process. Most eligible voters don't vote, even in general elections, and many don't even bother to register. They don't believe their vote really makes much difference one way or the other, so they just don't bother. First, they don't get to decide which candidates appear of the ballot. They can vote in primary elections, but a candidate can't be a contender in statewide primaries without a lot of corporate backing. Even if they like a candidate who is running, they know their single vote won't be of any real significance. The pollsters will have determined long before the election which candidates will win most of the races. The TV networks will call the close races, by "exit polls" on Election Day. Counting of votes is just a formality to validate what is already known. The real elections are won or lost in shaping of public opinion prior to the elections. So, why bother to vote?

Most don't. If only 60% of eligible voters actually register to vote and if only 60% of them actually show up to vote, a candidate can win an election with votes from less than 20% of the eligible voters actually voting for them. That's a "pretty scary" way to run a democracy! The corporations, not the citizens, are in control of the current American version of democracy. A "corporate democracy," quite simply, is not a rational form of government for any society. Yet few people are even willing to question why we have abandoned our democracy for corporatism.

My years of experience in university communities have provided direct, first-hand knowledge of corporatism in the public service sector. I worked at major Land Grant Universities in four different states over a period of 30 years. Over this time, I saw corporate influence grow from being so little as to be of no practical concern to so much as to threaten the very integrity of our public academic institutions. In the early days, in North Carolina and Oklahoma, corporate influence came mainly through influence on individual faculty members. Some faculty members did private consulting with corporations and came under the same type of influence as do honest politicians. Professors who spend a lot of time interacting with people who promote only one approach or paradigm in addressing public issues are unlikely to consider seriously any alternative paradigms in their research programs. These professors were not making laws, but they were deciding what kinds of research to carry out with taxpayer's money.

Over the years, corporations found ways to increase their influence on academics. During the Reagan years, society seemed to lose faith in its public academic institutions, along with its declining faith in govern-

ment in general. This loss of faith was not the product of a natural evolution in human thinking; it was the result of a thoughtfully devised corporate strategy to "get the government off their backs." The "best government is less government" attitude didn't stop with government regulations; it affected public attitudes toward everything the government did,
including public research and education.

Public funding for research and higher education began to decline,
or at least, to increase far more slowly than the universities' costs of
doing business. Universities were told that they would have to become
more efficient, that they would have to become more like private business. They were going to be held economically accountable for every
public dollar they received. The universities responded by doing what
was asked, they streamlined their educational assembly lines, and they
focused their research on things that could be measured, things that
would contribute to the economic bottom line. As they became more
"efficient" in doing things that could be counted, they became less "effective" in doing those things that truly served the public good. Although
the universities may have pleased the politicians and bureaucrats, they
were no longer pleasing the public. The academic quality of a college
education declined and important needs of the public, particularly the
needs of common people, were largely ignored in the public research
and education agenda.

Declining public support for public universities translated into
fewer tax dollars to support public research and education. By this time,
university administrators were thinking as corporate executives, and corporate executives equate downsizing with failure. It would take decades
to rebuild lost public confidence, so increasing public funding would be
difficult, if not impossible, in this era of privatization. The public universities turned to the private sector for help, or at least were more open to
private sector "offers of help." The universities were already accustomed
to receiving money from private foundations, trade associations, and corporate groups, presumably to support pursuit of the collective good of
their respective memberships. But by the 1990s, the public universities
began actively pursuing "big money" from corporations, many through
"joint ventures" with huge potential for private profits.

The Bayh-Dole Act of 1980 allowed public universities for the first
time to patent the results of federally funded research and for the universities to share in royalties with the corporations that commercialize their
discoveries. Numerous other laws have since been passed to tear down
the wall between public universities and private corporations – such as
giving generous tax credits to corporations that fund research at public
institutions. University administrators have rationalized such arrangements with the argument that the public can benefit from research dis-

coveries only after they have become commercialized and the products embodying the discoveries appear in the market place. They argue that joint ventures with private corporations focus research more clearly on commercial ideas and speed the process of commercialization.

Electronic information and biotechnology are big sources of private corporate funding for public universities these days. Both depend on the products of innovative, creative minds, which historically have been concentrated in the universities. Both may require large investments in expensive equipment that may become obsolete long before it is worn out. There are powerful incentives for private corporations in these areas to form joint ventures with public universities, to tap the brainpower of university faculties, and to focus public funds on the most risky aspects of the discovery process. But, there are also huge potential profits to be reaped from the discovery process.

So the corporations offer the universities an opportunity to stay on the cutting edge of research, more than offsetting their lost public funding with large private grants and contracts. Beyond that, they give the universities an opportunity to share in the potentially huge profits, and eventually to free themselves from the constraints of public funding and become a part of the private corporate sector they so much admire.

University administrators seem to have forgotten, assuming they ever understood, that true public benefits accrue to society as a whole, not to individuals, and that many public benefits cannot be valued in terms of dollars and cents. Certainly, there are public benefits associated with private goods and services, but there are private profit incentives to produce such things. Public funds are to be spent on those things that have inadequate, if any, profit incentives, and thus, will not be provided by the private sector, in spite of their potential benefits to the public in general. If the private sector has a profit incentive to carry out research, they will do it. Subsidizing research that private corporations eventually would do anyway, even without public funding, clearly is a misuse of public funds. No one really seems to care about the "common good" anymore, they rationalize, "what's good for Microsoft and Monsanto is good for America."

In addition to my early corporate and academic careers, I spent the last decade of my professional career working on programs that were funded by grants and contracts. Most were federal and state government contracts, but a few were grants from private foundations. I got the distinct impression that most of what I had observed in the corporatization of universities was true for government agencies and private foundations as well. Most public service institutions were being operated as corporate organizations; organizational growth and influence had become more important than pursuit of the "common good."

What does all this mean to the average person? It means that we, the people and citizens of this country, are no longer in control of either our economy or our government. We no longer have a capitalistic economy; instead, we have a corporatist economy. Capitalism simply does not work when business decisions are driven by corporate needs for profit and growth rather than by the uniquely human interests of individual investors. We no longer have a democratic society; instead, we have a corporatist society. Democracy simply does not work when political decisions are driven by the agendas of special interest groups, including business corporations, rather than by the uniquely human interests of individual citizens. The economic and political systems that made America great are no longer working – and won't work until people again make and take responsibility for their own decisions and actions.

Corporatism is the "specter haunting global society today" – not communism, socialism, or even true capitalism. Corporatism is the societal villain that threatens the future of humanity. The needs of the corporation can never be satisfied. The uncontrolled corporate society eventually will convert all human and natural resources into economic resources to meet its insatiable needs.

Industrialism is obsolete, capitalism is gone, and corporatism is obscene. Now is the time for dramatic and fundamental change, before it's too late. The world is continually changing. The invisible hand that once transformed greed into good has been mangled in the machinery of industrialization. Industrialization was the organizational paradigm of the past, but not of the future. Capitalism was the economic paradigm of the past, but not of the future. Corporatism is the societal paradigm of the present, but promises nothing but ecological and social destruction for the future. We must find the courage to confront these villains and to create new paradigms which can provide a desirable quality for life for ourselves and for all people of the future. We must find the courage to think and to act in new and different ways. It's time for new paradigms and new ideals. It's time for a return to common sense.

Endnotes

[1] *Merriam-Webster's Collegiate Dictionary*, 11th edition, "Corporatism" (Springfield, MA: Merriam-Webster Inc., 2003).

[2] Ray Anderson, *Mid-Course Correction: Toward a Sustainable Enterprise – The Interface Model* (White River Junction, VT: Chelsea Green Publishers, 1998) and *Our Wineries, Living in Good Taste*, <http://www.fetzer.com/fetzer/wineries/index.aspx> (accessed September 2006).

[3] John Elkington, *Cannibals With Forks: The Triple Bottom Line of 21st Century Business* (Stony Creek, CT: New Society Publishers, 1998).

6

AN UNCONSCIOUS SOCIETY:
THE CHALLENGE

After three years in Georgia, as a department head in a major state university, I decided that I didn't have much of a future in university administration. I don't think I was a particularly bad administrator, but I'm sure I wasn't the best. And more important, I had concluded this wasn't how I wanted to spend the rest of my career. I wanted to do something that truly contributed to the good of society; that's the reason I had decided to work for "public" universities. In Georgia, I noticed that faculty members in my department who were from Georgia seemed more committed to helping people work through their problems. Those of us from other states could always move away if we became frustrated. I wanted to feel that same sense of commitment in my work. I began looking for an opportunity to get back to the University of Missouri.

The USDA had just received an allocation of $3.8 million to initiate a new program that eventually was called "LISA," Low Input Sustainable Agriculture. My contacts at USDA told me the department "didn't have a clue" as to how it was going to spend the money. If I could come up with a good proposal that would productively employ my talents and abilities for a couple of years, it just might be my ticket back to the University of Missouri. MU wouldn't hire anyone who had received their Ph.D. degree from MU for a tenure track position – a common practice among major universities. But universities can always find places for people who can bring their own salary money with them. I developed a proposal, went to Washington to sell it to the folks at USDA, and before the end of 1988, my family and I were headed back to Missouri.

Through my work on "sustainable agriculture," I became aware of a newly emerging paradigm for economic resource development, called "sustainable development." That's also how I was introduced to the concept of "paradigms." A paradigm is simply a pattern or a mental model that describes how a process is organized and how it functions. Paradigms do three basic things. First, they define the goals and objec-

67

tives of a process – define the things to be accomplished in order to succeed. Second, they establish boundaries – define what is considered to be a part of the process and what is outside or external to the process. And finally, they spell out the rules – define what is and is not allowed in carrying out the process.[1] Games provide simple but useful examples of paradigms – tennis, baseball, football, golf, chess, and so on. All such games define what actions are required to score and to win, lay out clear boundaries within which the game must be played, and have specific rules by which the game is to be played.

In the real world, the goals, boundaries, and rules are more complex than in games, but the principles are the same. In the game of neoclassical capitalism, the goal is profits and growth. The business organization that continually makes profits and grows is deemed successful. The boundaries within which the game is played are defined by the social and physical or natural constraints. Things that are economic in nature are considered a part of the process, and things that are social or ecological in nature are considered as external or outside. The rules of the corporate societal game are defined by civil and criminal laws and by available technology. Anything that is legal and is technically possible is considered to be allowable.

The sustainable development paradigm is fundamentally different from the paradigm of neoclassical economic development. The goal of sustainable development is to sustain a desirable quality of life – economically, socially, and ecologically. To ensure sustainability, the environment, society, and economy must be in play – all three are within the boundaries of decision making. The rules of sustainability are the basic laws of nature, including human nature, which define the relationships within nature, including within society, necessary for sustainability. Sustainable economic development must be carried out in harmony with nature and society, regardless of what is considered legal or is technologically possible.

Most definitions of sustainable development are consistent with the definition adopted by a 1992 United Nations conference on Sustainable Economic Development. Sustainable development was defined as "development that meets the needs of the present generation while leaving opportunities for future generations to meet their needs as well."[2] To meet the needs of all people in the present without diminishing the opportunities for those of the future, development must be ecologically sound, economically viable, and socially responsible. All are necessary and sustainability requires nothing less.

Sustainable development is a common sense concept; it asks the question: Can development, meaning the productive use of natural and human resources, be sustained over the long run? Development is often considered to be synonymous with growth, but this narrow interpretation is based on the neoclassical economic concept of industrial development. Sustainable development simply means meeting the evolving needs of human society without diminishing the capacity of the natural and societal resources upon which humanity ultimately depends.

Sustainability is a matter of intergenerational equity; it applies the Golden Rule both within and across generations. To sustain humanity, we must care for other people, as we would have them care for us, as if we were they and they were we. We must do for those of future generations, as we would have them do for us, as if we were of their generation and they were of ours. The Golden Rule doesn't ask that we deny our self-interests, only that we show concern for others as well as ourselves. We must find ways to meet our needs today while leaving equal or better opportunities for others around us today and for those of the future.

The more I learned about sustainable agriculture, the more sense it made. It seemed to provide answers to many of my questions concerning whether my work as an economist was contributing something useful to society. Sustainability most certainly is related to economics, but it recognizes that economics alone is inadequate. Economics is important, but in the final analysis, everything is not "just a matter of economics."

Growing societal concerns for long run sustainability have arisen from a merging of the environmental and social justice movements. From the environmental movement, we have learned that we of this generation are using up the resources of the earth at rates far in excess of their rate of natural regeneration. Nature is nothing short of miraculous in its ability to resist our abuse, absorb our punishment, and regenerate and restore itself to health after we have exploited it. Until recent times, in fact, humans had not been capable of doing anything to the earth that it couldn't endure. We harvested its forests for shelter, mined the soil for food, and fouled its streams with our wastes. But we could always move on to other forests, fields, and streams; given time, nature would recover, regenerate, and renew itself. Today, however, as we continue our relentless degradation of nature, there are few unspoiled places on earth left to which we may move.

Some of our ecological sins can be attributed to ignorance. The word "environment" didn't even enter the vocabulary of most Americans until Rachel Carlson wrote her book, *Silent Spring*, in the early 1960s.[3] She wrote about the potentially devastating impacts of agricultural chemicals on wildlife and human life, and foresaw the day when there might be no birds left to sing. People had gained some awareness of the potential for humanity to destroy itself with nuclear energy during the 1950s. But nuclear annihilation was considered to be a product of war – not of day-to-day business. Somehow, we hadn't yet understood that we could destroy ourselves while trying to feed ourselves or trying to create any number of things that otherwise might make our lives better.

The energy crunch of the 1970s, when the Organization of Petroleum Exporting Countries (OPEC) effectively reduced world oil production, gave us our first glimpse of what life might be like in a world running short of fossil energy. U.S. petroleum production peaked in 1971 and has been declining ever since.[4] Today, evidence continues to mount that the world is approaching a peak in global petroleum production, after which supplies of petroleum inevitably will decline and prices will rise to ever-higher levels. All alternative sources of energy will be more costly and will create greater environmental risks.[5] The industrial era has been fueled by cheap energy and there are no sources of cheap energy left. Reliance on oil from coal, for example, the U.S.'s most abundant fossil energy resource, would present far greater risks of acid rain and global warming, which could be the first truly global environmental crisis to confront humanity. Eventually, all people must come to understand that when we pollute and deplete the resources of the earth, we are destroying our future. We are making the earth unfit for human life. There are no clean, fertile places left to "move on to," and yet we see exploding populations in "lesser-developed" nations and exploding per capita consumption in the "developed" and "rapidly developing" nations. We are now clearly capable of destroying our natural environment and thereby capable of destroying ourselves. We must now learn how to live sustainably upon the earth.

If human life on earth is to be sustainable, we must conserve nature's resources and protect the natural environment. It is not a matter of either limiting population growth or limiting consumption - both are inseparable parts of the same problematic whole. We must somehow bring consumption, population, and resource regeneration into balance to insure the well-being of ourselves, meaning all of us, as well as provide equal opportunities for those of future generations. Sustainable systems of development must be ecologically sound if they are to be socially responsible and economically viable over time.

From the social justice movement, we have learned that development based on the exploitation or degradation of people is no more sustainable than is development based on the exploitation or degradation of nature. Many people, even some who are vocal advocates for the environment, fail to recognize the importance of people in the development process. The fundamental purpose of all activities we call "development" is to improve the quality of life of people. Benefiting some at the expense of others is not sustainable. Sustainable development must enhance the quality of life for all - in some socially equitable fashion.

Starving people will eat their seed corn rather than save it to plant a crop, if they do not expect to survive to harvest. People who feel deprived of the actual necessities of life and are desperate will exploit their environment for short run survival, even if they realize they are putting future generations at risk. People in many parts of the world are turning their fields and forests into deserts as they deplete the fertility of their soil and turn trees into firewood in a daily struggle for existence. People who are chronically oppressed and are treated inequitably eventually rebel and destroy the societies that oppress them. An economy based on human exploitation is ever vulnerable to social revolution. Thus, a society that denies social justice to its people is not sustainable. Sustainable systems of economic development must be socially responsible if they are to remain ecologically sound and economically viable over time.

Sustainable development also must be economically viable. The economy provides the means by which people meet their wants and needs by relating to their natural environment and to each other in complex societies. If all people were self sufficient, they wouldn't need an economy. They would relate directly with nature and wouldn't need to relate to each other. But once people move beyond self-sufficiency, begin specializing and trading within their communities, and eventually begin to trade between communities, they need some means other than barter to facilitate efficient exchange. They also need some impersonal means of deciding who gets to use scarce natural resources and for what purposes they are used.

The economy – regardless of whether it is socialistic, communistic, or capitalistic – determines who gets to make the decisions regarding how resources are used. If any approach to resource development is not economically viable, it will not be used, regardless of how ecologically sound and socially responsible it might otherwise be. If it's not profitable, at least periodically, it's not sustainable. Sustainable systems must also be economically viable if they are to be ecologically sound and socially responsible over time.

Our corporatist systems of industrial economic development are polluting the natural environment, using up non-renewable resources, and otherwise degrading and depleting the resources upon which the future of humanity ultimately depends. Growing social inequities, disintegration of families and communities, and a general weakening of the social fabric of society are all symptoms of our corporatist system of industrial economic development. We are doing these things in the name of economic efficiency, but we are destroying and degrading opportunities to lead successful, productive lives – all for the sake of more cheap stuff.

As I began to awaken to the need for fundamental change, I was amazed that it had taken me so long to realize what was going on. Why had everyone, including myself, not seen this before? Then I began to understand. It wasn't obvious to everyone then, and isn't obvious to everyone now, because we are living in an unconscious society. Most people quite simply are not conscious of what is happening in the world around them; they are walking around as if in a dream. I have come to believe this is not a condition inherent to human society. This condition has been carefully developed and cultivated to perpetuate the ongoing process of corporate exploitation. This is not some evil conspiracy; it is just corporations doing what corporations are designed to do - make profits and grow.

People in our society are kept either working so hard or playing so hard that they don't have time to stop and think. We are just too busy and tired to pay too much attention to the ordinary things happening around us, unless there's a crisis. Working, and getting to and from work, takes a lot of time, as does all the other things we have come to believe are necessary. We simply don't have time or energy to be concerned about what's happening in the world around us.

Admittedly, the country has not been at war, at least not like earlier wars. People have not been rioting in the streets, and "nothing big" has happened to shake people out of their complacency. Even after the tragic terrorist attacks of September 11, 2001, the American people were urged to get back to "business as usual" as quickly as possible - most of all, to keep spending money. Most people did, and most have paid little attention since, even as soldiers and civilians alike continue to die. We were told that the attacks of 9/11 were a result of evil elsewhere in the world, so we had no reason to reexamine our own way of life. When the corporate accounting scandals of 2002 revealed some of the ugliness and destructiveness of corporate greed, we were told these were only a few "bad apples," that the integrity of the corporate economy, in general, was beyond reproach. Energy prices rise dramatically, oil companies make obscene profits, and we are told that's how free markets work. Americans remain oblivious to what is happening to their economy because they are told every day, in a multitude of different ways, that they have no reason to be concerned because the economy is fundamentally strong. The stock market may be up or down this week, or this year, but we are told that the long run trend is certain to be up. The vast majority of the people always have a job, a house, a car, and money to spend, and thus are better off financially than are most people in the world, and this is all that really matters.

Occasionally, someone will point out that we have an awfully lot of people in jail in this country. But crime has been worse in times past, and after all, prisons create jobs. Occasionally, someone will point out that we seem to have an awfully lot of people on drugs, but maybe that's because some people haven't yet learned to appreciate all of the things they have in this country. From time to time, someone will point out we are spending more money on doctors, hospitals, and medicine than we spend on food, but maybe that's because we have such great medical technology. After all, healthcare creates jobs. And at times, someone complains that we are spending a lot of money for lawyers and police, but maybe that's just because we have such a complex society. Law enforcement also creates jobs. And so what if the government has to pay to clean up a bit of industrial pollution now and them, even cleaning up after polluters creates jobs. The government really doesn't have anything better to do. There seems to be a convenient economic answer for every social and ecological question.

Those who criticize corporate exploitation of the environment and exploitation of people are labeled as "bleeding-heart, tree-hugging liberals." Liberals are just jealous because they aren't rich and aren't willing to work hard to become rich - like conservatives. If they don't like America, why don't they just go somewhere else, or else quite complaining about it?

The bottom line is that most Americans are afraid to complain about the economy. They are afraid they will appear ungrateful. After all, they live in one of the richest countries in the world. They are afraid they will be labeled as ignorant. After all, everything they hear and read tells them that they are blessed with the strongest economy in the world. They are afraid if they ask too many questions, someone or something will take what they have away from them. They think they had better just leave well enough alone.

Sometimes people do ask questions. Jimmy Carter, in his infamous "crisis of confidence" speech in 1979, openly questioned the sustainability of contemporary American society. "In a nation that was proud of hard work, strong families, close-knit communities, and our faith in God, too many of us now tend to worship self-indulgence and consumption. Human identity is no longer defined by what one does, but by what one owns. But we've discovered that owning things and consuming things does not satisfy our longing for meaning. We've learned that piling up material goods cannot fill the emptiness of lives which have no confidence or purpose."[6]

He called on Americans to restrain our greed in meeting the challenges of the energy crisis of that time - for the good of ourselves as well as for those of future generations. The next year, the American voters soundly rejected Carter at the polls. The people chose Ronald Reagan instead. Reagan reaffirmed the principle that "greed is good," and told us the thing most likely to create a crisis was a lack of confidence in our right to take as much as we could get. Reagan understood that this kind of confidence drives the demand for more and more stuff, and in turn, creates jobs and drives the economy to ever-higher levels. No serious contender for the office of President since has dared to hint of a "crisis in confidence" in America.

This "don't worry, be happy" theme of American public life has been carefully orchestrated by the corporations that make profits and grow as long as people – consumers, workers and investors – don't ask too many questions about the economy. Economic growth requires an insatiable consumer demand for more and more stuff. So people have to be convinced that having more stuff is the key to a more desirable quality of life. Thus, quality of life must be peddled to the American consumer as something they can buy. "If you aren't happy, it's because you don't yet have enough stuff." You just need to work harder so you can buy more; then you will be happy. The fact that the stuff you have hasn't made you happy yet just means that you don't yet have enough. Quality of life is measured by your standard of living and the American standards for the good life are always rising.

Those who point out that most of us already have too much stuff, and too little time, are a potential threat to the economy. We might just find we are actually better off spending more time building personal relationships and less time working; we would earn less, but we just might be happier. But the economy would suffer if we spent less. If too many of us spent more time with friends, family, or just by ourselves, and less time earning and spending, we just might create a recession.

Among the most senseless of all our American rituals comes each year at the winter holiday season. If we don't all go out and max out our credit cards, spending money we don't have, buying presents for other people that they don't need or even want, then the economy will fall into a recession. In other words, our economy is based on our spending money we don't have to buy things we don't need, and if we don't do it, a lot of people will be without work. We have to keep buying things so we will have a job to earn the money to pay the bills for the things we have already bought. Does that really make any sense?

Corporations also are caught up in self-perpetuating cycles of destructive behavior. If too many companies refuse to exploit their workers, the environment, or their customers, their costs of production will go up, their profits will go down, and the economy will suffer. Costs of raw materials, operating costs, wages and salaries, and employee benefits all will be higher than necessary, if a company exploits less than the laws allow. Widespread resource stewardship and compassion would add inflationary pressures to prices and recessionary pressures to the economy, meaning higher prices, reduced consumption, less employment, less growth. Corporations must keep their costs down, regardless of the costs to society. Does that really make any sense?

But, ordinary people are not supposed even to think about such things. We simply shouldn't worry, because worrying is bad for the economy. We can't afford to worry about other people because social welfare programs are a drag on the economy. If everyone would just learn to take care of him or herself, and expect everyone else to do the same, we would all be better off. Taxes supporting social programs could be better spent by taxpayers on things that contribute more to the economy. Even such popular programs as Social Security and Medicare are portrayed as dangerous steps toward Socialism – in spite of their undeniable popularity.

"Welfare" programs, such as Food Stamps, Aid to Dependent Children, and Medicaid, are clearly counterproductive, at least from a purely economic perspective. Certainly, all people should be given an opportunity to succeed, but opponents claim that we are creating a welfare state, a state in which a whole class of "losers" is becoming dependent on handouts from the winners. Social programs such as minimum wages, unemployment insurance, and workers compensation only succeed in raising labor costs to employers and eliminate jobs that might otherwise be available to willing workers. The economy thrives on low labor costs and the "rising economic tide lifts all boats, large and small."

Environmental programs are labeled as unwarranted infringement on private property rights as well as wasteful government spending. "Resource management" is said to be good for the economy – as long as public resources are managed to yield profits for private resource-based industries. "Preservationists" are a problem, because they want to keep natural resources in their natural state, and that deprives private industry of access to potentially productive resources. Resources are only of value if they provide jobs and income for people, which also require corporate profits, so we are told.

We are told we are foolish to be overly concerned about using non-renewable resources. New technologies are emerging all of the time. There is no limit to what we will be able to do in the future. We can find a substitute for any natural resource we use up – hydrogen fuel from water to replace fossil fuels, for example. We can learn to grow food by using hydroponics, or even in the ocean, if we allow our topsoil to erode away. And, industry is finding new ways to recycle and reuse waste all the time. We need not worry about resources for the future.

Corporate power in the marketplace is another thing about which people are told not to worry. The American economy is the strongest in the world, so our markets obviously are working. The economic concept of market structure – the number of firms in an industry and the share of the market they control – is an outdated concept, many economists now say. As long as we have three or four firms in an industry, we have enough competition, and even fewer is fine if the result is lower prices and continued innovation. We shouldn't worry about outsourcing good-paying American jobs to countries with lower labor costs; there will be plenty of jobs for everyone in the new global economy. Those who question free trade are isolationists; continued government intervention in trade threatens the future of the economy. Americans shouldn't be afraid to compete in the new global economy.

In general, this "don't worry" scenario portrays government as an obstacle to economic progress, and thus, an enemy of the people. If the corporations can make the people believe that the government is their enemy, then the people will side with the corporations against government. In reality, government is the potential enemy of the corporation, not of the people, because government intervention is the only means by which people can ensure that corporations serve the public interest. Whenever people distrust their government, the corporations are able to prevent the government from protecting the interests of the people.

Corporate advocates don't want common people to understand the proper role and function of government, because they don't want people to take control of their government, and through their government, to demand that corporations serve the public good. So corporations must convince people that they have nothing to worry about with respect to the private economy. If they worry at all, it should be about too much government interference with private enterprise, they are told. The people must be kept unconscious of the consequences of the growing power of corporatism.

At any point in time, American society may appear to be strong and healthy because the short run economic indicators are positive. In other words, if we consider only the usual economic risk factors, the probability of an economic catastrophe might seem negligible. But individual risk factors are often incapable of telling when something is fundamentally wrong with a system as a whole. I know this from personal experience.

Prior to my heart problems, I was in good physical condition. I had been jogging for 20 years, putting in 15-20 miles a week. My weight was normal and I hadn't smoked in 20 years. I drank a little wine, but only on the weekends, and wine was supposed to be good for me. My blood cholesterol was a bit high, but still within the normal range. After accounting for the positive risk factors associated with my healthy habits, my risk of a heart attack appeared to be negligible. But, there were internal stresses within an apparently healthy lifestyle. Those stresses on the system as a whole were slowly clogging my arteries and eating away at my health.

In the years just prior to my first visit to the hospital, my intuition was telling me something was wrong. It was becoming harder and harder to jog up and down the road. Eventually, I gave up, started walking, and just chalked it up to old age. I wanted to believe the positive risk factors. I told myself, "don't worry; just be happy."

Today, corporatism is eating away at the health of America – socially, ecologically, and economically. It is destroying the very things that ultimately must sustain the economy and society. America's natural resources and human relationships are under tremendous stress, as it becomes harder and harder to keep the economy growing without relying on increased ecological destruction and social exploitation. America's economic resources are being depleted by federal budget deficits and trade deficits, which mortgage the future to subsidize our greed. We are also depleting centuries of American good will through strategic wars to gain control of resources in other parts of the world.

American society, as a whole, is sick. Lacking some serious intervention, it is headed for a premature death. The American people know this intuitively, but are in denial. The people in positions of power and authority, the people they are supposed to be able to trust, are telling them that everything is okay. Jimmy Carter told the people the truth back in 1979, and look what happened to him. So Americans continue to search for individual short run risk factors that will tell them that everything is fine. They are unwilling to admit to the growing sense of societal illness that is gnawing at their hearts and souls.

We may have to be shaken out of our "don't worry, be happy" attitude by some sort of crisis – be it an economic collapse, global military conflict, or eruption of civil disorder. But we don't have to wait for such a catastrophe. We can break the bonds of corporatism, any time we choose; all we have to do is to reject the conventional wisdom and return to the first principles of human relationships – to common sense.

Our intellectual insight, our common sense, tells us we are working so long and hard that we have neither the time nor the energy to enjoy life. Happiness requires positive relationships with others and a sense of purpose and meaning in life. We know we can't consume our way to happiness. We don't really need someone else to tell us that jobs and income resulting from crime, sickness, and pollution are not good for society – no matter how much they contribute to economic growth. These things reflect sickness within our society, an inability to relate to each other and an unwillingness to fulfill our moral obligations. We understand that quality of life is about far more than economic standard of living.

A quiet and confident voice within tells us we cannot afford to ignore the welfare of other people, that a healthy society must be built upon the principles of equity and justice. We understand in the depths of our soul that we must do unto others, as we would have them do unto us – not just for their benefit, but also for our own sense of well-being. We have an innate understanding that the future of humanity depends on our taking care of the earth's resources and our natural environment. Caring for something for the benefit of others gives purpose and meaning to our lives. We don't have to be economists to know that economics should be about more than just quantity, price, and endless innovation – that an economy should meet the real needs of people. We must wake up from our "don't worry, be happy" hypnosis before it is too late. We must find the courage to listen to our common sense and break the psychological grip of corporatist propaganda.

History is filled with examples of people dominated for decades, if not centuries, by some oppressive regime – be it a warlord, chief, emperor, or central government – from which escape or restoration of personal freedom seemed impossible. Invariably, however, people find some way to break free from oppression. In medieval times, it must have seemed impossible that the serfs would ever free themselves from their feudal lords, but eventually they did. During the days of American slavery, I suspect most slaves had little hope of escaping from their owners, and their owners most certainly were not going to give up "their property" without a fight. Yet, the slaves were freed – even if their descendants are still struggling to achieve equality.

In America, we have the opportunity for "openness" – the constitutional freedom to criticize our economic and political systems. We have the opportunity for "restructuring" – the constitutional right to restructure our economic and political systems. We have the opportunity to free ourselves from the oppressive yoke of corporatism – we need only find the courage to return to and to trust our common sense.

Endnotes

[1] Joel Barker, *Paradigms: The Business of Discovering the Future* (New York: HarperCollins Publishing, Inc., 1993).

[2] *World Commission on Environment and Development, Our Common Future*, ed. Gro Bruntland (Oxford: Oxford University Press, 1987).

[3] Rachel L Carson, *Silent Spring* (Boston: Houghton Mifflin Company, 1962).

[4] *Wikipedia*, "Hubbert Peak Theory," <http://en.wikipedia.org/wiki/Hubbert_peak> (accessed September. 2006).

[5] Richard Heinberg, *The Party's Over: Oil War and the Fate of Industrial Societies* (Gabriola Island, BC, Canada: New Society Publishers, 2003).

[6] Jimmy Carter, "The Crisis in Confidence Speech," delivered July 16, 1979, *The American Experience*, Public Broadcasting System: Primary Sources, <http://www.pbs.org/wgbh/amex/carter/filmmore/ps_crisis.html>, (accessed September 2006).

7

A More Enlightened Self-Interest:
A NEW PERSONAL VISION

No revolution can succeed unless the people have a positive vision of an alternative future – a vision to which they are willing to commit their lives and their fortunes. This vision is already emerging in the hearts and minds of a growing number of ordinary people all across America and around the world who are marching under the various banners of sustainability. These people are pursuing a more-enlightened concept of self-interest that leads to a better quality of life for themselves and for others, both now and in the future. They are creating a shared vision of a better world. Through sharing with these thoughtful, caring people, I have come to see a new positive personal vision for the future.

As I began working on sustainable agriculture issues in Missouri, I began to realize that sustainability was about more than just an alternative paradigm for resource development; it was about a fundamentally different worldview. I began to see that concerns for sustainability arose from a different conception of how the world worked, and a different perception of the place of people within the global system. I began to realize that people who were concerned about sustainability saw a different world from those who saw no reason for concern. Thus, I concluded, the transition to a sustainable society will require a fundamental shift in the dominant global mindset or worldview.

The foundation for the dominant worldview was laid during a period in history that is sometimes called the "age of enlightenment." This period of human enlightenment, which spanned most of the eighteenth century, emerged from the scientific principles and laws developed by Newton, Descartes, and others during the seventeenth century.[1] This era is considered to have ended with the beginning of industrialization in the late 1700s. The French are credited with being the intellectual leaders during this time, with the most noted advocate being Voltaire, the French playwright, poet, and writer – although other Frenchmen did far more of the thinking than did he. The French philosopher, Charles de

Montesquieu, was one of the earliest advocates, but Jean Jacques Rousseau is credited with being the best thinker of the group. Germany's Immanuel Kant, Scotland's David Hume, and America's Benjamin Franklin, Thomas Jefferson, and Thomas Paine also were contributors to the critical thinking of this time.[2]

Perhaps, the most important assumption underlying the philosophies of that time was an unwavering belief in the power of human reasoning. The general belief was that God had made the universe, put it in motion, and then had stepped aside, leaving "man" to fend for "himself" in a mysterious world. The universe, including the earth and everything upon it, was believed to function according to a set of immutable, unchangeable laws that governed everything, including the natural ecosystem and human society. These rationalists believed that scientific inquiry and logical reasoning could solve the mysteries of the world, leading to unending human progress, not only in human understanding but also in technical achievements and even perfection of human moral values. "Man" didn't need to know God; "he" only needed to know God's laws. The new scientific methods eventually would unlock the mysteries of the universe, they proclaimed, revealing all of God's secrets. In essence, "man" then could be "his" own God.

The rationalists believed that true knowledge could come only through sensory experience and systematic observations of natural occurrences. They believed that humans could learn to manipulate nature for their own benefit, and that even human nature itself could be altered and improved through greater understanding and proper education. They accepted a Deist concept of God, the existence of some higher or natural order of things, but soundly rejected Christian theology. Human efforts should be centered on improving current life, not on some promise of life hereafter.

The scientific method, neoclassical economic thinking, and the industrial paradigm of economic development are all rooted in this age of enlightenment. The philosophies of this era probably made sense, given the state of human knowledge of the time, much as the economic principles of Adam Smith made sense in his day. However, the state of human knowledge and understanding has evolved a good bit since the 1700s. The mechanistic worldview proposed by Newton and Descartes no longer stands alone as the only logical hypothesis concerning how the world works. Einstein's theory of relativity and the various theories of quantum physics have seriously challenged the mechanistic worldview, and these so-called new theories can no longer be ignored. They reveal a world that functions much more like a living organism than a sophisticated machine.

The philosophies of the age of enlightenment no longer seem so enlightened. They may not be completely wrong, but they clearly are not adequate to address the pressing issues of today. The scientific method, a process of science, has supplanted the purpose of science, and that process no longer produces credible results. It's time for a more enlightened approach to acquiring and using knowledge, an approach based on a new set of first principles - a new foundation of common sense.

I am not suggesting that we simply substitute quantum physics for mechanical physics and chaos theory for statistical theory and then proceed anew to solve the mysteries of the universe. I am suggesting that scientific developments of the past century only serve to verify what common sense should have told us all along.

We should continue to seek answers to questions concerning how the world works, but we can never be certain that what we have found is the ultimate truth - even about the smallest of things. After each theory, there will always be another theory, and today's "facts" eventually will be replaced by tomorrow's new and better "facts." Each new theory and each new fact may improve our understanding of the universe, but we will never know all that we need to know to gain control over the world, or even over our own lives. We may increase our understanding of "how" the world works, but science will never bring us any closer to answering the questions of "why." We must continue to rely on our intelligent insight, our common sense, in answering questions that relate to the purpose and meaning of life - to do so truly is more enlightened. The purpose and meaning of life is determined at some higher level of organization, beyond the observable and measurable.

In many respects, the hypothesis of a Deist God still seems reasonable. It seems reasonable that some fundamental, immutable, unchangeable laws of nature, including human nature, govern the universe - including the earth and everything on it. But, it is not reasonable to believe that if humans were sufficiently intelligent or cunning, we could make the universe work according to our desires or preferences. The true laws of nature are inviolate. If human nature truly is a part of nature, it cannot be changed through greater understanding or education. It seems reasonable to believe that we must act in ways that conform to the laws of the universe or suffer the consequences of our violations.

It is reasonable to believe that we can continue to learn about the laws of nature, and thus can learn to reduce the conflicts in our lives and to live more in harmony with nature and with other people. But it is not reasonable to believe that humans are capable of fully understanding these laws, nor of using these laws somehow to manipulate the universe. By now, it should be obvious to all but the most egotistical that the working of the universe is far more intricate and complex than imagined by the eighteenth century philosophers. In fact, the universe could even be far more intricate and complex than scientists today can even imagine. Today's scientists are only beginning to scratch the surface of that complexity. Our intelligent insight tells us that we humans will never understand but a small fraction of all there is to know.

In the world of the 1800s, it may have been reasonable to discard superstitions and dogmatic religious beliefs in favor of a more science-based approach to understanding the functioning of the universe. However, time has proven that the enlightened thinking of two centuries ago is completely inadequate for the world of today. Today, it seems far more reasonable for humans to rely less on science and to rely more on their intellectual insight – our common sense.

Our common sense tells us that human beings will pursue their self-interests. I believe this to be one of the fundamental, immutable laws of human nature – a first principle of human behavior. If so, it cannot be changed, and thus we must reconcile ourselves to it, rather than waste time and energy attempting to change it. Common sense also tells us that a rational Creator would not have created a universe in which humans, by their very nature, would exploit the earth, and eventually destroy everything upon it. So why is our pursuit of self-interest degrading the natural environment and destroying the dignity and productivity of so many people? The answer is simple. We have been pursuing a wrong, unnatural concept of self-interest.

The contemporary economic concept of self-interest is derived directly from the philosophies of the age of enlightened reasoning. The "economic man" is assumed to be a "rational" being, in that "he" values only those things that arise from "his" sensory experiences. All economic benefits and costs, by assumption, are physical and tangible in nature. If we can't see, hear, taste, touch, or smell it, then it simply doesn't exist. The philosophy of rational materialism, upon which neoclassical economics is based, strongly denies any form or existence other than that of having some kind of palpable material characteristics and quality, which "stands in direct opposition to a belief in any of those existences which are vaguely classed as 'spiritual.'"[3] Consequently, the "economic man" values only those things that affect "him" individually and personally.

If the economic man does something that benefits another person, it is only because he expects to get something of greater economic value in return – something of tangible, palpable material value. He may choose to act in ways considered moral or ethical by the standards of his society, but he does so only because he expects to realize some tangible benefit or sensory pleasure, or to avoid some tangible cost or sensory pain. These experiences logically include productive work, which is necessary to free "man" from the uncomfortable constraints of a hostile environment and demanding society. Productive work provides both income and a sense of accomplishment, which allows "man" to pursue his self-interests.[4]

In the world of rational economics, one person cannot benefit from another person's success or happiness. Acts of true altruism are irrational. Since sensory experiences can only occur while a person is alive, no one can possibly benefit from anything that happens after his or her death. A person might do something for their children, or their grandchildren, because a child or a grandchild can return some sensory pleasure while the parent or grandparent is still alive. But, it is completely irrational to do anything for the sole benefit of those of future generations. There is simply no logical reason for such acts, since one cannot possibly reap a rational, tangible reward.

Fortunately, in many cases, the pursuit of one's narrow self-interest also results in benefits, rather than costs, to others, even though this was not the primary intent or motivation. Adam Smith observed that such relationships seemed to be the norm, rather than the exception, at least in his day. Even today, the economic man need not be a complete parasite on civil society, although in economics this is not a major concern one way or the other.

Of course, most economists don't really believe all of these things. A prominent agricultural economist supposedly once said that he certainly wouldn't want his daughter to marry an "economic man." Economists realize that most people are not heartless and soulless. I have known many good economists personally who were caring, responsible people. However, the pursuit of narrow, individual self-interest remains a cornerstone of contemporary, neoclassical economic theory. Economists are real people, but the "economic man" has no sense of human compassion or stewardship.

Our collective practice of rational economics has bought us to a logical destination. We are not a nation of purely "economic beings," but we have moved pretty far in that general direction. Many Americans tend to give freely of themselves only when it brings positive public recognition, when it salves their guilty conscience, or otherwise enhances their self-image. Americans are known for performing acts of great compassion and charity when confronted with local, national, or international emergencies or disasters. Some of these "selfless" responses obviously are motivated by the media publicity that surrounds such events. Others, however, are truly altruistic. A crisis "gives people permission" to be irrational, compassionate beings. They truly want to give, from the bottom of their heart, even if they "can't afford it." These people want to do something for someone else, even if there is no possible personal, tangible recognition or reward for doing so. They give because it enhances their social and spiritual quality of life to give, because it helps gives purpose and meaning to their life. Many would like to make such acts of caring and compassion a routine part of their everyday life, but they don't, because it would be seen as irrational and illogical.

Through our reasonable, rational pursuit of self-interests, we Americans have achieved material success such as the world has never before known. But in the process, we have systematically destroyed purely human relationships by converting them into economic relationships. And we have systematically destroyed any true sense of ethics and morality by converting them into economic assets. But still, our common sense tells us that quality of life is not a simple matter of tangible, sensory experiences – it is much more.

As humans, we will pursue our self-interest – simply because we are human. But, more and more of us have recognized that our "true" self interest is much broader than it has been defined by economic dogma. And, we've also recognized that humans can be misled into doing things that are contrary to this "true" self-interest, as is evident in all of today's industrialized societies. Our persistent, systematic pursuit of individualistic, tangible, sensory pleasure is diminishing our overall quality of life.

To reverse this destructive trend, we need to pursue the broader concept of self-interest, which recognizes that we human beings are inherently social beings. We need positive human relationships with other human beings to make our lives whole. We must learn to value the role of human relationships in enhancing our quality of life, regardless of whether these relationships result in any tangible, sensory rewards. We must learn to pursue a higher concept of self-interest, which recognizes that humans, by nature, are inherently spiritual beings. We need to live by ethical and moral principles in order to find purpose and meaning in our lives. We must learn to value the spiritual part of our being in enhancing quality of life, regardless of whether our ethical or moral acts result in any tangible, sensory rewards.

We need to pursue a more enlightened concept of self-interest – one that recognizes that humans are unique among species. Our uniqueness is not just a matter of our superior ability to reason, nor of our more rational pursuit of animalistic pleasures. Our uniqueness is our ability to make conscious, purposeful decisions that reflect an understanding that we are but a part of the far larger whole of all creation. We have the ability to understand that our own well-being is integrally related with a universe far larger than our world, a universe that spans both time and space. Each thing we do affects other people, which in turn affects the whole of society, and eventually comes back to affect us. Each thing we do has meaning and purpose within some higher order of things, across space and time, which in turn affects our quality of life. In our pursuit of a more enlightened self-interest, we must recognize the social and spiritual implications of everything that we do.

I am not attempting to start some new school of radical thinking. Such thoughts have been around for a long time. For example, Alex de Tocqueville, in his classic book, *Democracy in America*, wrote of a similar concept of self-interest in the early 1800's – he called it "self-interest rightly understood."[5] Tocqueville believed that survival of the American Democracy was critically dependent on deeply rooted religious beliefs, which constrained early Americans' pursuit of self-interests. He reasoned that if these strong religious beliefs were ever to erode, they would have to be replaced with a strong sense "that man serves himself in serving his fellow-creatures, and that his private interest is to do good." He wrote that early Americans believed strongly "that men ought to sacrifice themselves for their fellow-creatures... that such sacrifices are as necessary to him who imposes them upon himself as to him for whose sake they are made." Tocqueville believed that "self-interests rightly understood" reflected the fact that people benefit from fulfilling their proper role in the larger society in ways that could never be linked directly to one's narrowly-defined, individual self-interest.

Pursuit of a higher self-interest does not imply a return to mysticism and superstition. It is a sensible theory concerning the functioning of the universe that has been supported by philosophers, scientists, and common people, throughout the whole of recorded human history. Humans have practiced religion since the beginning of civilization. Newton, Descartes, Voltaire and Paine, all believed in the existence of a Deity. The early economists, including Adam Smith, gave great attention to issues of equity, justice, ethics, and morality. The neoclassical economists abandoned equity and ethics of classical economics around the turn of the twentieth-century, in their pursuit of scientific objectivity. And only in the last few decades have Americans been told, and apparently many have believed, that it is unreasonable and illogical to believe in anything that exists beyond our sensory experiences. Scientists have not and cannot prove that God does not exist. If God exists, if a higher order exists, it exists at a higher level, in the realm of pure knowledge, beyond the realm in which we humans can observe, experiment, and prove or disprove anything. Agnosticism and atheism are simply beliefs of convenience, in a society that has chosen to worship rationality and logic instead of God.

Pursuit of a broader self-interest is not the teachings of some radical cult movement. Throughout human history, people have sought out other people to form families, tribes, communities, nations, and civilized societies. It should be obvious to any observer, whether philosopher, scientist or ordinary person on the street, that people benefit from their relationships with other people. Certainly, some of these relationships are based on mutual sensory benefits – symbiotic, economic relationships. However, there is clear and compelling evidence that people need other people for reasons that are not sensory, tangible, or even individual, in nature. Love is one of the most thought-about, talked-about, written-about, and sought-after things in the world – and for good reason. Love may not actually make the world go around, but it makes the ride worthwhile. People who are "in love," and people who just love, do completely irrational, illogical things in their pursuit of their broader self-interests.

Our common sense tells us that humans are multidimensional. Throughout recorded history, people have referred to the body, the mind, and the soul. Our body is our physical being, our mind is our thinking being, and our soul is our spiritual being. The physical, mental, and spiritual aspects of humans are not separate parts, but instead are inseparable dimensions of a whole person. Like height, length, and width of a box, if we take away the physical, mental, or spiritual, we are left with something other than a human. An object without height is a flat surface, and without height or width, is nothing – certainly not a box. Likewise, a being without a body, mind, and soul is not really a human being.

A being without a body obviously is not a person. Neither is a being without a mind a person, as we explicitly recognize when we "unplug" a body from life support because it is "brain dead." And, a being without a soul is not a person – a fact that has yet to be explicitly recognized by many in contemporary American society.

Since people are multidimensional, it should be no mystery that the quality of a person's life is likewise multidimensional. The tangible, sensory aspects of our life are fundamentally physical in nature. Thus, our individual personal quality of life, which includes the quality arising from economic self-interests, is related to the physical dimension of our being. Values that arise from human relationships are fundamentally mental and emotional in nature. The sense of well-being that arises from positive relationships with another being is not something we can see, touch, hear, taste or smell, but is clearly something that affects our mental and emotional well-being. Human values that arise from ethical and moral behavior are fundamentally spiritual in nature. The sense of well-being that arises from "doing the right thing" is not something tangible that can be sensed physically or often even something that we can rationalize mentally. It is something we experience deep within our soul. The personal, relational, and spiritual dimensions are inseparable, and together, determine our overall quality of life.

My older brother, Tom, claims that all human problems would be solved if we only "put the soul in control." He argues that our refusal to listen to and to follow what our soul tells us is not only the root of all evil but also the basic source of every human problem. In one sense, I agree with him, but I don't think it is quite that simple. I believe God gave humans a body and a mind for some positive reason – not just as temptations for the soul. I agree that people should listen to their soul; that's how we access our common sense. But I also believe that we are expected to think about what we are doing and to heed our physical instincts. If we were solely spiritual beings, there would be no purpose for the mind and body. To function as purposeful whole people, I believe we must function in ways that result in harmony among the physical, mental, and spiritual – among the body, mind, and soul.

This concept of harmony is consistent with religious beliefs. For example, a scribe once asked Jesus which of the Ten Commandments was the most important. Jesus answered "the first... you shall love the Lord your God with all your heart, with all your soul, with all your mind, and all your strength. The second... you shall love your neighbor as yourself."[6] Jesus said we are to love our God with all of our being – our body, our mind, and our soul. And He quickly added that we should love our neighbor as ourselves – that we should love both our neighbors and ourselves. I believe also that Jesus would admonish us to include those of future generations among "our neighbors."

These general admonitions are a part of virtually every enduring, organized religion. All religions are based on a belief in a higher order of things – in some form of God, even if only the inviolate laws of nature. All religions also have some version of the Golden Rule – to treat others, as we would want to be treated if we were they. In many enduring philosophies, we find the reference to three dimensions of life: the physical, the mental, and the spiritual; the personal, the interpersonal, and the intergenerational; the body, the mind, and the soul.

Economics has focused solely on the physical, personal, material dimension of our lives. In reality, the American preoccupation with economics has become a "religion" – in every sense of the word. It has a specific worldview, a well-defined value system, a set of guiding principles, a set of institutions devoted to its promotion, a discipline and group of scholars dedicated to keeping and protecting its principles, and a host of devoted disciples – including some very powerful, non-human organizations. Even our national holidays and religious celebrations have become economic events, marked by extravagant spending. But like any religion, this worship of economic self-interest is based on beliefs, not facts.

Economics as religion is a very recent concept in terms of human history. Its classical conceptual roots go back only a couple of centuries; its current neoclassical version is only about a hundred years old and has been in widespread practice for only a few decades. It has not stood the test of time, over thousands of years, as have many of the other religions of the world.

Today, the foundational beliefs of the religion of economics are crumbling. As we reject the beliefs of neoclassical economics, we must not revert to religious dogma. The dogma – the long lists of shalls and shall-nots – is no better than the dogma of economics. We must instead return to the first principles upon which all true religions and all civil societies must be built. Accepting the fundamental social and ethical first principles from past centuries is no more regressive than is accepting that the earth revolves around the sun or the laws of gravity. First principles are neither old nor new – they never change. We need a new age of enlightenment so we can build upon time-tested principles of the past to create a better human civilization for the future. We need to come out of the darkness of economic oppression and into the light of a new era of human progress. To make this great transformation, we must return to our common sense.

A society built upon a foundation of enlightened self-interest offers the best hope for a sustainable future. The three cornerstones of sustainability – economic viability, social responsibility, and ecological integrity – are related directly to the three dimensions of self. Economic viability is necessary for sustaining the personal, physical self. Social responsibility is a reflection of the necessity of sustaining the interpersonal, mental and emotional self. And to sustain ecological and social integrity over time, we must necessarily express our ethical, spiritual self. We must give conscious, purposeful consideration to the economic, social, and ecological implications of everything we do.

Some argue that one or another of the dimensions of sustainability is more important than are the others, just as my brother argues that if we put the soul in control everything else will fall into place. Some people argue that relationships are all important, that love alone can make the world go around. Some people argue that economic viability is most important, that if something is profitable, it's sustainable, period. Some people argue that ecological integrity is the most important – that the ecological laws of nature cannot be changed, and thus, we must adjust our economic system and social values to accommodate the natural ecosystem.

It's true that human society is but one element of the global ecosystem and that the economy is but one element of human society. In a hierarchical sense, the ecosystem contains the other two systems. However, hierarchy does not necessarily imply superiority or authority. It's obvious that people, a part of the natural ecosystem, now have the ability to seriously damage the natural ecosystem, if not destroy it. Ecological integrity alone will not ensure long run sustainability. It's also true that profits are a necessary part of economic viability, and economic viability is a necessary dimension of sustainability. But profitability is not a sufficient condition for long run sustainability. Without a productive resource base and a civil society, there is no economic viability. Profitability can be sustained only through ecological and social integrity. Social responsibility is just as necessary for sustainability as the other two, since a society ultimately decides what type of economy it will support and whether or not it will even try to protect the natural environment. Social relationships are important, but positive personal relationships alone can't sustain or ensure the integrity of either the ecosystem or the economy. None of the three is any more or less critical to sustainability than are the other two. All three are necessary and none alone is sufficient.

The key to a sustainable quality of life is balance and harmony, for individuals, for communities, for society as a whole. There may be times in our lives when it makes sense to spend more time and effort on one dimension than on the others. For example, it may make sense to spend more time working, when we are building a career, more time at home, when our children are small, or more time contemplating spiritual matters, when dealing with a mid-life crisis. But the stages of life must somehow fit together, in balance and harmony, to sustain a lifetime of well-being and happiness. Inevitably, stresses will arise among the physical, relational, and spiritual aspects of our lives, but stress among the parts only serves to strengthen the whole, as long as stress is not allowed to degenerate into distress. The key is balance and harmony.

To achieve a life of harmony, we must bring all dimensions of our lives together. When we live our life in little airtight compartments – never bringing our work home, never taking our home life to the office, never taking the office to church, or taking church to the office or home – we don't have a chance. We are whole people and we must be willing to live whole lives, not parts of lives that we somehow cobble together. We cannot make up for a deficit in one aspect of our life by achieving more in another. Life is interconnected. We can't do one thing without affecting everything else. Everything we do affects the balance and harmony of our lives. We have to work on it all together, at the same time.

The same principles apply at all levels of organization and aggregation. Sustainable families, communities, business organizations, and nations all require balance and harmony among the three fundamental dimensions: personal, interpersonal, and ethical – economic, social, and ecological. But, sustainability must begin at the personal level.

We don't have to be philosophers, religious scholars, or rocket scientists to understand these things. The problems and promises of sustainability are all matters of intelligent insight – of common sense. During
the age of enlightenment, people were admonished to abandon the mysticism and superstitions of the past and to rely on scientific reasoning
instead. It made sense at the time. But reliance on rationality and scientific reasoning has been taken too far – it has become an end in itself,
rather than a means to an end. Everything that is not physical, tangible,
or sensory has been discarded as being unreal, or at least as irrelevant.
Rationality and scientific reasoning most certainly have important roles
in human well-being, but our common sense tells us that other things are
important as well – things that we cannot see, hear, touch, taste, or smell.
For those things, we must learn to rely on our common sense.

When I talk about a great transformation, I'm not talking about going
back to the Dark Ages. I'm talking about moving forward. Today, our common sense tells us that our lives have important physical, mental, and
spiritual needs that must be met to achieve a desirable quality of life. Our
common sense tells us that we need balance and harmony among these
dimensions of our lives. Common sense also tells us that we can have a
sustainable human society, but that it has to begin with us – at the personal level. The vision for a fundamentally better future is clear and compelling – it is worthy of the commitment of our lives and our fortunes.
We have the power to shape our own destinies. We have the power to
realize the vision of a better world. To make the transition from an
exploitative past to a sustainable future we must first return to our common sense.

Endnotes

[1] *Microsoft Encarta Encyclopedia*, 2003, "Age of Enlightenment" (Redmond, WA:
Microsoft Corp., 1993-2003).

[2] Thomas Paine, "Age of Reason," in *The Life and Major Works of Thomas
Paine*, edited by Philip S. Foner (New York: The Citadel Press; 1961; republished, 2000 by Replica Press, Bridgewater, NJ).

[3] Hugh Elliott, "Materialism," in *Readings in Philosophy*, eds. John Randall, Jr.,
Jestus Buchler, and Evelyn Shirk (New York Harper & Row Publishers, Inc.,
1972), 309-310.

[4] Ayn Rand, *The Virtues of Selfishness* (New York: Penguin Books Inc., 1961-64).

[5] Alexis de Tocqueville, *Democracy in America* (New York: Bantam Books, 2000,
original copyright, 1835), 646-649.

[6] *Holy Bible, Revised Standard Version*, Mark 12:30-31 (New York: The World
Publishing Company, 1962).

8

LIVING ORGANIZATIONS:
A NEW ORGANIZATIONAL VISION

I have always considered myself to be a scientist. A scientist typically is defined as someone having expert knowledge in some particular
field of science, such as physics, chemistry, biology, and geology.
Economists generally are referred to in academia as social scientists, but
nonetheless, as scientists. In a more general sense, science refers to systematic knowledge, consisting of verified laws and tested propositions
presented as an orderly system. A scientist then is one who seeks knowledge or truth in some systematic, orderly fashion. Contrary to popular
belief, there is no single systematic, orderly means of gaining knowledge
or seeking truth. The appropriate method of scientific inquiry depends
upon the nature of the knowledge and truth one is seeking.

Likewise, there is no single systematic means of forming or managing an organization. The appropriate means of forming and managing an
organization depends upon the nature of the purpose of the organization. To organize means to arrange or form into a coherent unit or functional whole - to arrange elements into a whole of interdependent
parts.[1] Thus, any orderly, functional, interdependent arrangement of
things is an organization - including businesses, non-profit groups,
churches, political parties, governments of all types, and even factories,
machines, communities, and families. One of the most fundamental
threats to the sustainability of human society today is the inappropriate
use of the old mechanistic paradigm of industrial organization.

Thankfully, in response to this threat, a new organizational paradigm
is emerging that reflects a more enlightened worldview and vision for
the future. The new organizational paradigm is based on the concept of
living organisms - on dynamic, regenerative, self-making systems. The old
mechanistic worldview and industrial organizational paradigm have
become exploitative and degrading of nature and of people. The new
organismic worldview supports a sustainable organizational paradigm,
which is nurturing and regenerative of both society and nature.

The industrial organizational paradigm is based on a preconceived notion that the world works like a big, sophisticated, complex machine. Industrialization has its roots in the early seventeenth century when Rene Descartes laid the philosophical foundation for the dominant scientific worldview of today. Descartes believed that God had created a world made up of two classes of substances. One class he called "extended" substances, which included all material things or physical matter; the other was "thinking" substances, which included the mind.[2] He considered living things, specifically humans and some other animals, to be made up of both "thinking" and "extended" substances. Descartes believed that all non-living things, even the "non-thinking parts" of living things, operated as if they were sophisticated machines. He frequently used the metaphor of a clock, with its many precisely interrelated parts, as an analogy for how the "non-thinking" world worked. Each part has a specific function and a specific spatial and sequential relationship to all other parts.

Isaac Newton, an Englishman, built upon the work of Descartes and others to develop the mathematical foundation for contemporary scientific thinking. Although best known for deriving the law of gravity, Newton created the concepts of calculus and discovered many of the fundamental laws of mechanical physics. Newton, like Descartes before him and most scientists since, viewed the world as being fundamentally mechanistic in nature.

Therefore, according to modern scientific thinking, the world works like a big machine. All we have to do to understand it is to take it apart, piece by piece, and observe how the parts work together within the whole. The parts are separable – so we can understand the whole of a thing by understanding the functioning of its individual parts. That's what the term "scientific analysis" means: taking things apart and examining the pieces. In a mechanistic world, all functions can be defined in terms of causes and effects; each occurrence has a specific identifiable cause. Thus in modern science, each phenomena of interest is either the cause of or the effect of something else.

Scientists generally begin their inquiry by making general observations of relationships among things and then formulate hypotheses regarding the specific cause-effect nature of the phenomena of interest. Next, they design experiments, or make additional specific observations, designed to either verify or refute their hypothesis. Finally, they draw some logical conclusion regarding the validity of their hypothesized relationships, based on the results of their experiments or observations. This is the "scientific method" or scientific approach of understanding the nature of reality.

In the world of science, reality exists in terms of facts – in terms of objectively observable, verifiable, replicable, tangible experiences. Our senses are often aided by sophisticated sensing devices – microscopes, telescopes, listening devices, feeling devices, picturing devices – however, in science, nothing is considered to be real or true until it is verified by some tangible process. For example, scientists have developed various hypotheses about "black holes" in space. But, such hypotheses achieve scientific credibility only when scientists "see something" in space that seems to validate their claims. The purpose of science is to discover and define reality, and reality can be verified only through sensory experience.

This all sounded logical and reasonable until a few decades ago – but not today. In the early twentieth century, the world of mechanical physics was shaken at its very foundation. Albert Einstein put forth his "theory of relativity," Werner Heisenberg proposed his "uncertainty principle," and a handful of other scientists developed new "quantum" theories of physics. Einstein showed that concepts of time and space are not absolute, but relative. Heisenberg showed that the position and momentum of subatomic particles could not be measured simultaneously, and thus reality was not always observable. But quantum theory, in particular, turned the world of classical mechanical physics upside down and shattered the philosophical foundation that had supported modern science.

Defenders of the scientific status quo staunchly argue that quantum theory only applies to things at the subatomic level – for things so small that most of us don't even realize they exist. However, the most advanced theories of physics – including "string theory" and "brane theory" – appear to resolve the apparent conflicts between classical physics and quantum physics.[3] Thus, things that are true at the subatomic level may quite likely be true of all higher levels of organization as well. Quantum theory has clearly shown that at the subatomic level the whole of reality is interconnected. This suggests that all phenomena at all levels of organization are interconnected; and thus, it is impossible to isolate causes and effects completely. Each cause has many effects, which have many other causes as well, and which in turn become causes of still more effects.

Quantum theory suggests that the properties of objects are not independent of their environment. Reality is determined as much by the interconnections among things as by the things that are connected, and all things are interconnected. In quantum theory, reality exists as "potentials," which become "real to us" only when the phenomena of interest is observed in a specific context – in relation to other things. Things

"become real," when we observe them, and thus, we can't observe anything without changing it from potential to reality. And, since a thing
might be observed by more than one person in more than one context,
things have more than one "reality." So things we call "facts" are a function of human observation – of human consciousness.

A classic quandary in quantum reality is a question regarding the "cat
in the box." As the story goes, there is a cat in a box. There's no means of
determining whether the cat is alive or dead without opening the box.
Question: "Without opening the box: tell me, is the cat alive, or is it
dead?" The answer, "it's neither alive nor dead – not until you open the
box and discover it to be one or the other." In reality, it has the "potential" of being either one or the other. Such is the nature of quantum reality. And on the surface, such a concept of reality seems completely illogical and irrational.

A now-classic experiment in quantum physics pertains to the nature
of light. When scientists devised a method for observing light as energy,
as light waves, they were able to observe it as energy. But, when they
devised an alternative method to observe light as matter, as particles or
photons, they were able to observe it as matter. In mechanical physics,
everything is either energy or matter, but not both. In quantum physics,
light has the "potential" to be both; you find the reality you look for. Our
assumptions concerning the nature of phenomena determine our methods of observation, and thus, shape the nature of the phenomena we
observe.

Science, in general, is based on the proposition that reality is objectively observable, replicable, and verifiable. However, quantum reality is
not objectively observable, replicable, or verifiable; it changes from time
to time and from observer to observer, depending on the aspect of its
potential that is observed. Quantum reality depends on the conditions
under which the observation is made, including the means of observation and person observing.

So how have modern scientists been able to observe the same phenomena, replicate experiments, and verify each other's results, generally
coming to the same conclusions? The answer, even though everything is
interconnected, some interconnections are much weaker than are others. For "non-living things," many of the connections apparently are quite
weak, and thus, cause and effect relationships can be effectively isolated
and observed. The ranges of potentials within non-living reality apparently are quite limited and similar. In living systems, however, the interconnections often are strong and important, making isolation of cause and
effect virtually impossible. Every part of a living plant or animal is potentially critical to the life of the organism. Relationships among people
both within and among families, communities, and nations are potential-

ly critical to the quality of life of the people involved. The ranges of potentials within living realities apparently are quite wide and diverse.

The "hard sciences," such as chemistry and physics, deal with non-living things – with chemicals, minerals, gasses, etc. – where connections are weak. Even the researcher is less likely to be biased by his or her previous experiences with chemicals and minerals than by previous relationships with people, or even with live animals. For non-living things, a mechanical approach to science has been extremely useful and has done no great harm. For science related to living organisms – such as botany and zoology – the results have been impressive, but have left many important questions unanswered. But for sciences related to living systems, such as economics, sociology, and ecology, where relationships are strong, the mechanical approach to science has been largely ineffective and frequently destructive. There are no "mechanical laws" of economics, sociology, or ecology. In complex living systems, whatever is observed is always significantly dependent upon the conditions under which the observation is made and observer who made it.

Apparently, Descartes was at least partially right; "God did create a world of reality made up of two classes of substances" – the living and the non-living. A mechanistic approach to science may be acceptable, if not completely accurate, for dealing with non-living things, as Descartes suggested. But a new approach is required for dealing with the all-important questions of living things – including thinking things, as Descartes suggested.

Chaos theory is another twentieth-century scientific development that has shaken the foundations of predictive science. A Frenchman, Jules Henri Poincaré, supposedly anticipated chaos theory in the late 1800s when he questioned the predictability of motion in the solar system.[4] But Edward Lorenz, an American meteorologist, is credited with developing the modern theory of chaotic motion. Lorenz showed that weather models, although based on fairly simple mathematical equations, were extremely sensitive to initial assumptions and to measurements of weather conditions. He concluded that it was scientifically impossible to make accurate long-range weather forecasts, although clearly definable weather patterns made short run forecasting quite feasible. The behavior of weather patterns over time is chaotic, he claimed.

Many natural systems are simply too complex to accurately model and forecast very far into the future. Natural phenomena often have too many complex and significant interrelations. Typical examples of chaotic phenomena include smoke, steam, clouds, streams, ocean currents, disease outbreaks, insect infestations, and plant and animal populations. I would quickly add nearly all cultural, economic, and sociological phenomena to this list. Errors in forecasting such phenomena are not "random errors," as assumed in statistical forecast models. Instead, the errors are the consequence of chaotic complexity. High-speed computers have allowed scientists to show that general patterns will emerge out of long-term chaos – given a sufficient number, like billions, of observations. There is definite order within the chaos. The shorter the timeframe, the fewer the interactions, and the more predictable will be the pattern. But as the timeframe is extended, it becomes impossible to predict where any future observation will occur within that pattern. There are general patterns or trends within even the most complex reality, but the nature of specific occurrences remains fundamentally uncertain.

I first became interested in chaos theory when I was trying to forecast livestock prices in Oklahoma. I have since concluded that most markets, including the stock markets, are chaotic phenomena and thus are inherently unpredictable over any significant period of time. Market analysts who successfully discern short-term patterns and trends in prices can create an illusion of predictability. But markets can shift chaotically and dramatically, to a higher or lower level, catching the best of market experts totally by surprise. I've seen it happen many times and it has happened to me. Economists simply cannot forecast the future because the economy is chaotic.

During my last few years in Oklahoma, I discovered that month-to-month changes in cattle prices were being influenced far more by cyclical changes in the spread between prices of beef in retail stores and live cattle prices, than by changes in fundamental supply and demand conditions. The cycle I discovered was about 11 months. I had discovered a short run pattern in a market that I knew, from previous research, to be fundamentally unpredictable over the long run. There were broad patterns or cycles in cattle supplies and prices; but the long run cycles were too variable in length and amplitude to be of any real use in long run price forecasting.

As long as my short run pattern held, I was forecasting cattle prices with a great deal of accuracy. But I knew from my previous research that spreads between retail-beef and live cattle prices were not predictable over time. They were determined in a dynamic system with a host of interdependent variables, including many and varied expectations and lagged reactions. The causes of changes in price spreads, as in market prices, are too many to be isolated and analyzed using conventional scientific methods. The interdependencies are too strong and the processes too complex to predict with any degree of accuracy. I knew once something happened to break this particular pattern, I wouldn't have a clue as to what type of pattern might emerge next.

I began to see this same type of phenomena in all types of living organizations, including economies, societies, and ecosystems. They all seem to behave in an orderly, predictable fashion for a time, but then something seems to trigger a dramatic, unanticipated change – a recession, a revolution, or an ecological collapse. Such times of apparent chaos eventually are followed by times of comparative order and predictability. Living systems in general are not predictable over any extended period. Living systems do function according to basic underlying principles, and thus order exists within the chaos, even for complex living systems. But if science is to be of any real benefit in managing economics, societies, natural ecosystems, or in any other living system, science itself will have to change.

Chaos theory has no obvious relationship to quantum physics, except that both challenge the mechanistic worldview that has dominated thinking throughout the modern scientific era. The chaotic behavior of observable phenomena, such as weather and markets, has definite similarities to quantum behavior of subatomic phenomena, although I am not aware of anyone attempting to link one to the other scientifically. Both most certainly raise questions about the adequacy of a mechanistic, deterministic science as the foundation for contemporary thinking.

The important conclusion here is that a mechanistic science may be adequate for dealing with non-living systems, but we need a fundamentally different approach to science to deal with issues involving living systems. The differences between living and non-living systems are subtle, at least in the abstract, but apparently are important in understanding the whole of reality.

Fritjof Capra, a physicist, contends that all systems, living and non-living, possess three basic characteristics: pattern, structure, and process.[5] Pattern is the conceptual framework for the system. For a non-living system, the pattern is the blueprint or design. For a living system, the pattern is encoded in its DNA – in its genetic code. For both living and non-living systems the conceptual framework is constant, unchanging, and remains fixed over time. A bicycle always is a bicycle, for example, and a person always is a person.

The structure of a system is the physical manifestation of the pattern. For non-living systems, the structure is the thing you see or touch – the bicycle, automobile, or building. For a living system, the structure also is the thing you see or touch – the plant, animal, or human body. However, the primary difference between non-living and living systems is found in the dynamic nature of living structures. For non-living systems, the structure is fixed – it can never change on its own. It may wear out and it may be rebuilt or redesigned, but it has no ability to change itself. A machine keeps its same physical structure for all of its useful life – a bicycle always has the same size, shape, and form. However, the structure of a living system is in a continual state of change and renewal.

Living things are born, they grow, they mature, they reproduce, and they die – the baby, the adult, and the old man are all the same person, but their physical characteristics are always changing, and over time, change dramatically. This continual change in structure is a fundamental characteristic of all living things.

The nature of process also is different for dead and living systems. Non-living systems achieve their purpose by performing a linear sequence of functional processes: input, transformation, and output. The fundamental purpose of non-living systems is to transform the energy potential of an input into a more useful form of energy output. A person rides a bicycle to transform kinetic energy stored in leg muscles into mechanical energy that turns the wheels and propels the bike down the road, taking the rider from one place to another. An engine transforms the kinetic energy in a fossil fuel into the mechanical energy needed to perform some useful task. Input is transformed into output.

Living systems perform useful purposes as well, but living processes are also self-renewing and self-regenerating, as well as functional. Living systems devote a significant portion of their energy to replacing the cells of their bodies, thus transforming their physical structures in the process of performing their productive functions. As the human body transforms kinetic energy in the process of performing useful physical and mental work, it also builds muscles and bones, changes physically, grows, and reproduces, before it eventually dies in the process of living. Living processes are circular and simultaneous rather than linear and sequential. As plants grow, they convert solar energy into protein, carbohydrates, and fats – energy forms that can provide energy for animals or for the next generation of plants. Function and regeneration occur simultaneously for living systems – they renew themselves in the process of fulfilling their purpose.

In summary, non-living systems are designed to accomplish some purpose according to some blueprint or pattern, their structure is fixed or constant, they function for the duration of their usefulness, and then they must be rebuilt, redesigned, or discarded. For living systems, on the other hand, the pattern and purpose is embedded in their genetic make-up, in their DNA. The processes of a living system include both functional usefulness and self-renewal. Living systems continually change and renew their structure in accordance with the unchanging genetic code embedded in their DNA. Living systems also are capable of evolving from generation to generation, on their own, to accommodate their ever-changing environments.

So what difference could these new ways of thinking – quantum physics, chaos theory, and living systems – have on the way we organize things? First, most human organizations today are structured according to the old mechanistic worldview – appropriate only for non-living systems. The result is a world filled with exploitative, unsustainable industrial organizations, because they devote nothing to regeneration and renewal for the benefit of future generations. Second, an alternative organizational paradigm, based on the principles of living systems, could fill the world with human-friendly, earth-friendly, sustainable organizations, which evolve from generation to generation to meet the changing needs of human society.

Dee Hock, the founder of Visa Corporation, has developed what he calls a "chaordic" organizational paradigm as a replacement for the old industrial organization. Hock claims "the industrial age, hierarchical command-and-control institutions that, over the past four hundred years, have grown to dominate our commercial, political, and social lives are increasingly irrelevant in the face of the exploding diversity and complexity of society worldwide. They are... increasingly unable to achieve the purpose for which they were created, yet continu[e] to expand as they devour resources, decimate the earth, and demean humanity. We are experiencing a global epidemic of institutional failure that knows no bounds."[6]

In one respect, all human organizations are the same, industrial or otherwise. Any organization, no matter what its structure, must have a purpose; otherwise, there is no logical reason for bringing people, money, and other resources together. However, for an industrial organization, the purpose is embodied in the organizational design. An industrial organization is designed so that its specific functions, procedures, and responsibilities, if carried out properly, will ensure that the purpose of the organization is achieved.

A well-run industrial organization works like a well-oiled machine. Each machine is designed to fulfill a purpose – which may be as simple as drilling a hole or as complex as assembling the body of an automobile. Each part of a machine is designed to perform a specific function by a specific process so that all parts work together, allowing the machine to fulfill its purpose. Each machine is controlled by an operator who may do something as simple as flipping a switch or as complex as guiding the machine through a series of intricate maneuvers. However, the role of the operator is matched with the design of the machine; together they perform a specific function.

A machine must be maintained if it is to continue to perform effectively. A poorly maintained machine breaks down too often and wears out too soon. Even under the best of care, individual parts eventually wear out and have to be replaced. Machines with interchangeable, replaceable parts can be repaired rather than replaced, and thus, have a tremendous advantage over machines that are manufactured as single units. Eventually however, any machine will become obsolete – it will no longer be able to fulfill its purpose, at least not as well as some newer design. Eventually any machine must be either redesigned or discarded and replaced with a newer model.

An industrial organization, like a machine, is designed for a specific purpose. Each position in the organizational chart, from chief executive officer to production line worker, is defined so as to perform a specific function in furthering the purpose of the organization. An industrial organization requires constant maintenance to ensure that each person in the organization performs his or her function in support of the overall organizational purpose. Even in the best of organizations, individuals eventually "wear out." Workers eventually become disabled, retire, or they simply lose their commitment or usefulness to the organization, and they have to be replaced. However, a "new person" can always be hired to fulfill the specific responsibilities of the "old person," and the organization will continue to function as before. In industrial organizations, the people are simply interchangeable parts.

If the industrial organization becomes obsolete – is unable to perform its purpose as effectively as some competitive organization – it must be reorganized, restructured, or redesigned so as to make it run more effectively. The ultimate responsibility for redesign lies with those who own the organization, the stockholders in the case of a corporation, but this responsibility typically is delegated to top-level management. Regardless, someone must decide when an organization has become obsolete and thus must be redesigned or discarded.

A chaordic organization, like any organization, must have a purpose. However, the purpose is much more prominent and important in the chaordic organizational paradigm. In an industrial organization, the purpose is of primary importance at the design stage – the purpose is designed into the structure. In the chaordic organization, the purpose must be instilled in the hearts and minds of every person in the organization. In living organizations, the focus must be on the commitment of the people who fill the positions, rather than position descriptions, and the people must remain personally committed to the purpose of the organization.

The essence of this new living organization is embodied in its principles of operation, rather than its organizational structure. The principles of a living organization are embodied in a set of standards for individual and collective conduct. The guiding principles must be both necessary and sufficient to ensure that the organization fulfills its purpose. If a principle is not really necessary, it will unduly constrain the ability of the organization to adapt and change in order to continue fulfilling its purpose. If the set of guiding principles is not adequate or appropriate, the organization may not function effectively in pursuing its purpose.

The structure of a living organization is dynamic rather than fixed. Positions, departments, divisions, organizational units, take on new meaning over time. They are continually changing and evolving, forming and dissolving, as the organization transforms and renews itself to meet the ever-changing demands of a dynamic marketplace in an ever-changing economic, social, and natural environment. This is the chaotic part of Hock's chaordic organizational model. The order part of the chaord is embodied in the set of organizational principles, which guides its processes or functions. The purpose and principles of the organization, the conceptual framework, remain unchanging, while the structure evolves as needed to maintain the effectiveness and efficiency of the organization. The principles of a living organization only evolve from generation to generation, as necessary to accommodate changes in organizational purpose.

For the people who work in living organizations, the principles are fundamentally different from the functions that make up a position description. A person in a living organization may still have specific responsibilities, but he or she is free to meet these by any means consistent with the principles of the organization. The person in the position, not the position description, determines the most appropriate means of pursuing the purpose of the organization. And the person may choose to change their means of fulfilling their responsibilities, in order to adapt to different situations or changing organizational environments. Thus, the focus of living organizations is on purpose, principles, and people.

During my last five years at the University of Missouri, I provided leadership for a three-state sustainable rural development project, which had been funded by the W. K. Kellogg Foundation. For five years, the project team attempted to carry the project according to the principles of a "living organization." We built the project organization around a fundamental purpose and a set of guiding principles. Individuals were free to pursue a wide range of means to contribute to the overall purpose of the project, as long as they remained true to the fundamental principles.

The experiment was a successful learning experience. At least I know I learned a lot. For example, it was difficult to explain to those involved how a "living organization" is meant to function because we are all so accustomed to the industrial model of organization. I had even more difficulty convincing people that we could carry out a successful project without the usual rigid organizational structure spelling out the hierarchical chain of command and control or a specific set of rules and operational procedures. People kept asking what I wanted them to do and how they should go about it.

The best example of a living organization I was able to find as an example of a living organization was that of an effective democracy. The conceptual framework of a democratic republic is encoded in the Constitution. The purpose in forming the Union, the purpose for the organization, was spelled out in preamble to the U.S. Constitution. "We the People of the United States, in order to form a more perfect Union, establish justice, insure domestic tranquility, provide for the common defense, promote the general welfare, and secure the blessings of liberty to ourselves and our posterity, do ordain and establish this Constitution for the United States of America."

The fundamental principles by which the Union was to function were spelled out in various Articles of the Constitution and in The Bill of Rights. The first three Articles of the Constitution define the structure of the government - with its legislative, executive, and judicial branches. Article IV addresses relationships between state and federal levels of government. Article V makes it clear that the Constitution was to be a living document, in that it makes clear provisions for the people to amend and even rewrite the Constitution, whenever deemed necessary to meet the changing needs of the people.

Our experiment with a living organizational paradigm in Missouri was less than completely successful, primarily because we were unable to instill and maintain a commitment to the purpose and principles of the project. However, I remain convinced that a living organization not only provides a far more hospitable climate in which to work, but is potentially far more productive than is the more common industrial organization. The people who work in organizations are living beings, and much of the truly important work being done today relates to living organizations - families, communities, economies, and societies. It is just common sense that our human organizations should be modeled after living, thinking systems. By organizing around principles, the structure of living organizations can continually change and evolve as needed to continue fulfilling their purposes. The living organization can empower people to use their uniquely human capacity to think and to act on their own to help create a better world.

Most important, the new living organization can be self-sustaining. Current concerns for the sustainability of development are rooted in some of the most basic laws of nature. For example, the first law of thermodynamics, generally referred to as the law of conservation of mass and energy, might suggest that sustainability is ensured. Einstein's theory of energy conservation - $E=mc^2$ - states that mass and energy are equivalent. Mass may be converted to energy and energy converted to mass, but energy and mass in total are conserved, and the total thus remains unchanged.

On the other hand, the second law of thermodynamics states that whenever energy is transformed in the process of performing work -

mass converted to energy, or energy changed in form – some of its "usefulness" is lost. This loss in usefulness results from energy changing from more concentrated and more organized forms to less concentrated and less organized forms in the process of doing work, as when gasoline explodes in the cylinder of an automobile or water is transformed into steam to drive a generator. In order to reuse the energy, it must be reorganized, reconcentrated, and restored, all of which requires energy, which then is no longer available to do work. As their total amounts of useful energy are depleted, closed systems inevitably tend toward entropy, "the ultimate state reached in degradation of matter and energy; a state of inert uniformity of component elements; absence of form, pattern, hierarchy, or differentiation."[7] This second law of thermodynamics, the law of entropy, might suggest that entropy is inevitable and sustainability is impossible.

However, the first and second laws of thermodynamics relate to "closed systems" – where nothing is lost to the "outside" and nothing comes in from the "outside." With closed systems, which include all non-living systems, entropy is inevitable. Thus, long run sustainability of life on earth is possible only because of the openness of the earth to the inflow of energy from the sun and the ability of living things to capture it. Sustainable development is possible only because the earth, as an open system, is capable of capturing and storing sufficient amounts of useful solar energy to offset the declining usefulness associated with the inevitable tendency toward entropy.

This dependence on solar energy implies that sustainable development is dependent upon living systems. Living organisms capture energy from the sun, concentrate it, organize and store it in more diverse and useful forms. Living systems thus have the capability of offsetting the inevitable degradation of energy and matter. The natural tendency of living systems is toward greater diversity in structure, form, hierarchy, and pattern – away from entropy. Scientists have developed synthetic solar collectors of various types, and energy captured from wind and water is indirect solar energy, but if these systems are to be sustainable, they must function according to the principles of living systems. Living systems – plants, animals, humans – must provide the organizational paradigm for the sustainable use of all forms of solar energy. Sustainable development is inherently dependent on the organizational paradigm of living systems.

A sustainable society must conserve, recycle, and reuse materials and energy, if it is to slow, rather than accelerate, the unavoidable process of entropy. And ultimately, human population and per capita consumption must accommodate the carrying capacity of the earth. But, global carrying capacity depends at least as much on our ability to capture and store solar energy for use by future generations, as on our efficient use of the stocks of energy with which the earth is endowed. Solar energy can be used directly for productive purposes, but a portion of whatever is captured must be concentrated and stored for future benefit. Sustainable development must be a self-renewing, regenerative, living process.

Contrasts of living and non-living systems, of sustainability and entropy, are equally relevant to cultural, political, and economic systems. We humans are part of the regenerative, self-renewing living culture that has been passed on from generation to generation. Industrial organizations, be they institutions, economies, or societies, are based on a mechanistic paradigm, and thus, inherently tend toward entropy – toward degradation of resources and destruction of life in the process of extracting usefulness from energy. Sustainable systems must nurture and renew life and must encourage human imagination and creativity, if people are to live and work in harmony with nature.

The processes of sustainable systems must be both productive and self-renewing, if they are to meet the needs of present and future generations. Our human organizational structures, likewise, must be self-renewing and regenerative if they are to change and evolve as necessary to remain productive in an ever-changing natural and cultural world. The current industrial organizational paradigm is a root cause of the depletion of natural resources and degradation of human society. Our current organizations – businesses, government institutions, social structures – are not sustainable. A new organizational paradigm is emerging and eventually must become the dominant paradigm if our society is to be sustainable. This new paradigm is based on the first principles of living systems. This new organizational vision is our hope for a sustainable future. We only need the courage to throw off the old and take on the new. It's time for a great transformation, beginning with a return to common sense.

Endnotes

[1] *Merriam-Webster's Collegiate Dictionary*, 11th edition, "Organize" (Springfield, MA: Merriam-Webster Inc., 2003).

[2] *World Book,* 2002 Standard Edition, "Rene Descartes" (Chicago: World Book Inc., 2001).

[3] Brian Green, *The Fabric of the Cosmos; Space, Time, and the Texture of Reality* (New York: Random House, Inc., 2004)

[4] *Microsoft Encarta Encyclopedia*, 2003, "Chaos Theory" (Redmond, WA: Microsoft Corp., 1993-2003).

[5] Fritjof Capra, *The Web of Life* (New York: Random House Publishers 1996).

[6] Dee Hock, *Birth of the Chaordic Age* (San Francisco: Barrett-Koehler Publishers, Inc. 1999), 5-6.

[7] *Merriam-Webster's Collegiate Dictionary*, 11th edition, "Entropy" (Springfield, MA: Merriam-Webster Inc., 2003).

9

GOVERNMENT FOR THE COMMON GOOD:
A NEW SOCIETAL VISION

When I took on the administrative responsibilities for a department at the University of Georgia, I didn't want to be a full-time administrator. I wanted to maintain my academic expertise in Agricultural Economics. As department head, I had the authority to decide what type of educational program I would conduct. I decided to specialize in agricultural policy. Department heads were expected to interact with the leaders of the various agencies and organizations involved in agricultural policy, and I had always had an interest in public policy issues, so it seemed a good fit with my new position.

People tend to think of government farm programs as being programs designed to help farmers. However, legitimate government programs must serve a greater public purpose, rather than just serve a particular sector of the economy. The original intent of farm programs, which were initiated during the Great Depression of the '30s, was to stabilize food production by stabilizing prices for farm commodities. During hard economic times, when prices are well below costs of production, farmers are forced to cut back on production, which eventually causes prices to rise. But agricultural production is a biological process; it takes time to bring new production to the market and production processes cannot be easily stopped once they have begun. So, long delays typically occur between drops in farm prices and initial reductions in supplies and the eventual rise in prices. The result is chronically reoccurring periods of scarcity and surplus, which creates inefficiency in food markets and inconvenience for consumers.

When government farm programs were initiated in the 1930s, about one-fourth of all Americans lived on farms.[1] By stabilizing farm prices at near break-even levels, government programs could not only reduce the financial stress of low prices on farm families but could also create significant benefits for food consumers. By helping to keep family farmers in business, government programs would also benefit all residents of farming communities, not just the farmers. Farm programs were justified because they were good for society as a whole, not simply because they provided financial support for farmers.

I understood that government had a legitimate role to play in society in general, not just in agriculture. I had been taught that government was the means by which people pursued their collective interests – the means by which we can serve our individual interests better by acting together. Farm policy was just one example of this. Today however, most mainstream economists seem to believe the primary mission of government is to ensure uninterrupted economic growth of the private economy. That's why we see so many advocating privatizing virtually every public service that is still being provided by government. Beyond facilitating economic growth, economists consider government's primary role to be that of dealing with market failures, and few economists are willing to admit to very many situations where the markets have failed. Most seemed to believe there are few things that the markets can't do well – or at least better than the government.

While working on policy issues in Georgia, I eventually came to believe the lack of understanding of the legitimate role of government was a major problem confronting American society. My interest in public policy continued after I returned to the University of Missouri. My department offered courses in agricultural policy, but these courses dealt with policy from an historical and institutional perspective with little attention to the legitimate functions of government. The Economics Department offered courses in public policy, but again the emphasis was on how government policies worked and not why we needed them. So, I designed a course that would address public goods and services from the perspective of the legitimate role of government in serving the public good.

I worked with a group of rural sociologists to integrate the proposed course into a new curriculum. It was accepted, but unfortunately, too few students enrolled in the course to allow me to teach it. As far as I know, there is still no course offered at the University of Missouri that deals with the legitimate role of government in a civilized society. However, I still believe that participating in such a course, or some similar learning process, is essential to a good education, not just for college students, but also for any member of a democratic society.

A true society is something more than a collection of individuals. A society is a whole. The nature of the relationships among people within a society is as important as the nature of its individual members. A root cause of many of the problems confronting industrial societies everywhere is our inability to maintain strong positive relationships. Perhaps nowhere in the world is this problem more serious, and of more global significance, than in the United States. One of our fundamental problems is our current interpretation of the legitimate role of government.

Americans seem to have lost any real sense of the inherent value of societal connectedness, and thus, see little value in government. We may come together in times of crisis, but under most normal circumstances, we seem to prefer going it alone. In his book, *Bowling Alone*, Harvard University political scientist, Robert Putnam provides measure after measure indicating the extent to which Americans have become socially disconnected over the past fifty years; most measures of social connectedness dropping by 30-50 percent.[2] We vote less often, we attend fewer public meetings, and when we do, we are disappointed to find that few of our neighbors have joined us. We remain interested and critical spectators of the public scene, but we don't participate in public affairs. We remain affiliated with various civic associations, but we don't show up. We are less generous with our time and money, we are less likely to give strangers the benefit of the doubt, and they return the favor, according to Putnam.

We seemingly have abandoned the idea that it's necessary for us to work together for our common good. Even if Adam Smith's invisible hand were capable of translating our greed into good, which it is not, it would serve only our individual interest, collectively. Even perfectly competitive markets do not build positive human relationships, and thus, do not serve the common good of a society. The common good is something more than the collective total of individual goods, just as a commonwealth is something more than the mere collection of wealth.

The new societal vision for a sustainable future is a vision in which people understand that they must live and work together for their common good. Government is but one means, although an important means, for pursuing this new vision. For the most part, living and working together is an interpersonal matter – something to be worked out among people, one-on-one. As we learn to appreciate the contribution of personal relationships to our quality of life, we will willingly spend more time and effort building such relationships. Beyond purely personal relationships, all sorts of civic associations provide us with opportunities to work together on matters of community and local interest. However, for many relationships, particularly at the state and national levels of society, we must learn to work together through government.

"The government that governs least governs best." This has been a commonly held view in the United States since its beginning. The Founding Fathers were understandably skeptical of the power of big government, after their unfortunate experiences with the British Monarchy. In fact, the Bill of Rights of the U.S. Constitution is devoted primarily to ensuring that the rights of citizens are protected against governmental abuse. Skepticism regarding the legitimate powers of government is often considered a cornerstone of American democracy.

However, the American Declaration of Independence suggests otherwise. "We hold these truths to be self-evident, that all men are created equal, that they are endowed by their Creator with certain unalienable rights, that among these are life, liberty, and the pursuit of happiness," it begins. And it continues, "That to secure these rights, Governments are instituted among men, deriving their just powers from the consent of the governed." It is obvious that the founders of the American democracy believed that all people have certain rights that are undeniable and are equal for all people, which include the pursuit of happiness as well as life and liberty. It is also clear that a fundamental purpose in forming a government was "to secure these rights."

Earlier versions of the Declaration had included the phrase "life, liberty, and possession of private property." However, the Founding Fathers apparently concluded, quite wisely, the opportunity to possess private property was not equivalent to the pursuit of happiness. The United States Constitution spells out the fundamental purposes of our government: "to establish justice, insure domestic tranquility, provide for the common defense, promote the general welfare, and secure the blessings of liberty to ourselves and our posterity." Clearly, this was not a government designed to govern least, but instead, to govern best. People who don't believe there is any legitimate role for government other than to protect private property and promote the private economy are pursuing a philosophy of government fundamentally different from that envisioned by the Founding Fathers of the American Democracy.

The new societal vision for sustainable American society is but a new understanding of what the original American vision of democracy was intended to be. The new vision for the U.S. government is that of defender of the common good rather than simply the defender of private property.

The fundamental purpose of government is to ensure social equity and justice for all. All of the necessary functions of government spelled out in the Constitution stem from this basic purpose – including national defense, law enforcement, general welfare, and civil liberties. Assurance of equity and justice is the best means of ensuring the ability of all to acquire and secure private property and to benefit economically, as well as socially and spiritually.

Pursuit of common good does not imply government ownership of all property or even a significant amount of property. Common good does not imply Communism, but instead simply recognizes the common purpose of people in any society. The new societal vision recognizes that private property rights are both necessary and appropriate in those many cases where our interests are clearly individual in nature. There is no disagreement with the Founding Fathers who felt that protection of private property was essential to the welfare of the nation. But, the protection of the common good requires recognition that many societal values accrue to the people in common, and not to just individuals. Thus, we must pursue those common values together.

Our pursuit of the common good must be rooted in the realization that value arises from our personal relationships and from our sense of purpose and meaning in life, as well as from our individual self-interests. The quality of our lives is affected by the way we treat other people, and the way we feel about ourselves, not just by the amount of personal property we can accumulate. Certainly, the government must protect property rights; but it also has equally important roles in protecting the values that arise from human relationships and from our common virtue, values that arise from our being part of an equitable and just society.

The U.S. democracy, by design, allows us to pursue the common good through government, but it doesn't necessarily force us to do so. The U.S. Constitution clearly spells out the rights of citizens to conduct business through the private market economy, including exchange of private property. However, the Constitution also clearly spells out that all citizens have certain rights that are held equally by all, without regard to their ownership of property or anything of economic value. These rights are not to be bought and sold in the private market place; these rights are to be assured equally to all, regardless of what they have to offer for sale or are willing to pay.

If you and I go into a Wal-Mart store, you with a hundred dollars while I have only ten, you have the right to buy ten times as much stuff as the I do. Wal Mart is a legitimately private business. However, if you and I go into a voting booth, you with a hundred dollars and I with only ten, we each have only one vote. The voting booth is the place where people make public decisions, where all people have equal rights. In a democracy, we vote on matters affecting the common good, and we each have one vote, regardless of how wealthy or poor we may be. Each person has an equal voice in making public decisions because each person has an equal right to benefit from public goods and services.

Through government, we accept the responsibility to secure the rights of life, liberty, and the pursuit of happiness equally to all people. The private market economy only gives people rights in relation to their wealth or ability to pay. And we are inherently unequal in aptitude and ability to contribute to the private economy, and thus to make money and accumulate wealth. Thus ensuring the right to own and exchange personal property most certainly will not alone ensure the realization of those rights that, according to the Constitution, are to be equally accessible to all.

The government must ensure equality where equality is a right and ensure the rights of private property and free markets where equality is neither necessary nor desirable. The private sector has a legitimate and important function in rewarding people for their creativity and productivity, and thus, providing an incentive to earn money and accumulate wealth. A viable private sector is necessary for a strong economy. But, the public sector also has a legitimate and important role in ensuring equal opportunities, regardless of one's ability to earn or to pay. A viable public sector is necessary for a strong society.

Questions concerning which goods and services should be available to all, and thus, which are legitimate public goods and services, are questions that ultimately must be answered by society, by the people, either directly or through their elected representatives. However, some rights to public goods and services are promised by the Constitution, and thus, cannot be denied, short of a constitutional amendment.

For example, the Constitution ensures that all people have a right to be defended against foreign aggression. And, all people living in America have an equal right to be defended, regardless of whether they pay any taxes or have any property to defend. It seems only fair that those with more property pay more for national defense, but all have an equal right to personal protection, no matter how much or how little they pay.

All Americans have a right to move freely about the nation. So, public transportation – including highways, roads, and bridges – is considered a legitimate public service. In some cases, we interpret "equal access" to mean "at a cost everyone can afford," rather than "available free of charge." We do have toll roads, where speed and convenience is important to some drivers, and where alternative routes are available for those who are unable or choose not to pay the toll. But, most seem to agree that some form of public transportation is a legitimate public service.

Most Americans also agree that some level of education is a legitimate public service. We don't all agree on what level and type of public education is necessary, nor is there universal agreement on how best to provide public education. But most agree that an educated society is essential to maintaining an effective democracy and a productive economy, and that we all benefit when everyone is "educated," regardless of their ability to pay.

Social Security Insurance is another popular public service program. The first government funded "old age pension" programs were established during the Great Depression, when old people were actually starving and dying from exposure and hunger. The people of this nation concluded it was intolerable to live in a society where old people are allowed to suffer and die, just because they hadn't accumulated enough wealth during their working years to support themselves in retirement.

Those not eligible for Social Security and are old and poor are covered by other social welfare programs for the aged. Economic security for the old has strong political support as a legitimate public service.

Public support for social welfare programs in general is not as clear and strong as for public food programs or assistance for the aged. Yet most people seem to believe that all Americans have a right to some minimal level of food, and that children, in particular, should have sufficient nutrition to ensure their physical and mental development. The USDA Food Stamp program and various state and local social welfare programs are targeted to making sure all have access to sufficient food to keep them alive, and hopefully, healthy. Private markets provide food only for those who can afford to pay for it. Therefore, ensuring that everyone has access to enough food to survive is a necessary and legitimate public service. So, while most seem to agree that everyone is entitled to some minimal level of government support, if they truly need it to survive, there is no consensus concerning the level of support necessary or the conditions under which survival becomes the responsibility of government rather than the responsibility of the individual.

Another whole class of public goods and services currently exists that I will call "collective purchases," which are things that we choose to buy together, rather than separately, just because it is more practical to do so. These are not public in the sense that we all have an equal right to them, but only in the sense that government is the most practical, least cost, most convenient, or only feasible way to obtain them. Economics, in general, considers nearly all public goods and services to be "collective purchases" because economics simply doesn't deal with issues of equity and justice.

The government may be the least-cost means of securing goods and services when significant savings can be realized through large-scale purchases. This situation logically exists for many inherently public goods and services; so many such services are purchased collectively as simply a matter of practicality. For example, it isn't very practical for people to use vouchers or tax credits to buy their own tank or missile, to build little pieces of highways, or hire a teacher part-time to educate their children. However, we choose to make many other purchases collectively, such as electrical power, communications systems, water and sewer lines, parks, and recreation facilities even though they may not be inherently public in nature. Such services become equally accessible to the public, not because we all have an inherent right to equal access, but because they were bought with public tax dollars. We all may have equal access to such things, even if we don't pay equal taxes, because it is impractical to exclude those who didn't pay taxes or to limit those of us who paid less.

"Natural monopolies" have represented a common economic justification for collective purchases in the past. Natural monopolies included such things as electrical power lines, telephone and telegraph lines, railroads and highways, sewer and water lines – typically, situations where costs of building the infrastructure is very high in relation to the value of the service provided. It is simply impractical to run three or four electric power or cable lines to every house, to build two or three parallel railroad beds, or to run a half dozen different water or sewer lines all around town. Naturally, the company that built the first one of any of these things would have a monopoly. No one else could afford to build another with the promise of only half the market. In the case of natural monopolies, the market begins and ends with only one supplier.

In such cases, it was logical for the government to intervene on behalf of the public. The government would build the necessary infrastructure to provide the service or would grant a private company the right to provide the service, but then would regulate the quality and price of the service. Since the company granted the right would have a monopoly position in the market, there would be no competition to ensure quality service or a competitive price. In recent years, economists seem to be far less concerned about monopoly power. The emphasis today seems to be on privatizing virtually everything, without the requisite concomitant regulation, regardless of the obvious monopoly power granted to the private companies that provide such services.

In addition to providing public goods, the government also is responsible for protecting all people equally against the public bads. The people of a democracy have a right to "equal protection under the law." Most of these rights are generally understood, and most people accept equal protection as a legitimate function of government. Many of our moral and ethical principles are encoded into criminal and civil laws. For example, there is a national consensus that it is wrong to commit murder or assault and to steal or commit fraud, so we have laws that punish such acts.

In general, we have laws against "sins," but only if they affect other people. It is not against the law, for example, to commit adultery, to lie about personal matters, or to hate another person. Most would consider such things to be morally wrong, but they are considered to be personal matters to be resolved among individuals. Where criminal laws exist, society has reached a consensus that a victim has a fundamental right to be protected from specific unethical or immoral acts of others.

Some may ask why is it necessary to have laws if society has already reached a consensus concerning whether something is right or wrong. Laws are necessary because a consensus is never complete; some few people will never choose to abide, or perhaps are incapable of abiding, by the ethical or moral values defined by a consensus of the societies in which they live. Those who have worked for and have reached consensus must be protected from those few who refuse to participate in the civil processes of self-governance.

We have finally reached a consensus in America that everyone has "civil rights" – rights to be treated equally, as an individual, without regard to the specific group, or groups, with which they might be identified. Therefore, everyone has a right to be protected, in all public matters, against discrimination based on their race, gender, ethnicity, age, physical ability, or sexual orientation. Individuals still have the right to discriminate against people as distinct individuals. We don't have to treat every member of the public the same, or even to do business with everyone. But, we can't discriminate against individuals just because they are members of a particular group of people. The government has the responsibility to enforce our national consensus against discrimination.

The new societal vision for the future is a vision of people taking an even more proactive approach to working together for the common good through government. Anytime a more proactive role for government is suggested, the question arises as to how we going to pay for it. Each thing a government does must be supported by taxes. People always seem to feel that they already paying too much of their income in taxes. "It's my money, and I want to decide how to spend it," is the typical response to proposals for new government initiatives or programs.

In a democracy, the people, through their elected representatives, have every right to decide what functions they want their government to perform. They have a "right to decide how their money will be spent," but they also have a responsibility to decide such things together, not just individually. People have a right to decide together how much taxes they will pay and what kinds of public services government will provide. But, they don't have a right to decide they want the benefits of government programs but are not willing to pay the taxes necessary to support them. Taxes are compulsory contributions of money made by the people who are governed to support the things they must do together. And the people must be willing to impose the taxes upon themselves, not individually, but together.

In general, systems of taxation should be simple, straightforward, and sensible – they should make sense. First, everyone must realize that virtually all taxes are paid in dollars and cents; so all taxes are paid with funds derived from the private economy, by one means or another. Whether paid by a corporation, a partnership, or an individual, money must be earned in the private economy before taxes can be paid. The question is who will be taxed and how much they will be taxed on what they earn, to support the common good.

Tax collections should be linked as closely as possible with the government goods and services for which they are collected – this just makes sense. It's fairly easy in cases of collective purchases to link the public good or service with something that can be taxed. For example, electrical power, communications systems, and water and sewer lines can be supported by taxes paid by those who benefit most directly from the service, as is generally the case today. Gasoline and motor-fuel taxes are legitimately used to help pay for highways, bridges, airports and other public transportation services. But, some portion of transportation services is meant to be truly public services, ensuring the freedom of all to move from place to place, and thus the full cost of transportation need not be borne only by users of the service.

The costs of freely accessible collectively purchased services can be assessed to those who benefit economically from having such services available. Property taxes might be the logical choice to support parks and recreation facilities, for example, as the availability of such amenities invariably enhances local property values. Public education has also been financed largely through property taxes, under the premise that local schools benefit local residents economically, as well as socially.

The costs of inherent public programs, those designed to ensure equity and justice for all, should be shared broadly across the whole economy. It seems logical that national defense, law enforcement, education, and health care might fall most clearly in the category of programs to which all have equal rights. So it might be reasonable to support the costs of such services through a value-added tax – a tax assessed as a percentage of the gross domestic product accounted for at each stage of production. A similar tax might be levied on services. The broadest, most inclusive functions of government should be funded through the broadest, most inclusive form of taxes, such as value-added taxes.

Personal income taxes could be reserved to support those public goods and services that are most easily addressed by redistributing income. Clearly, we need to rethink the whole issue of income taxes – including why and how. The current system is so complex that no ordinary person can fully understand the extent to which it redistributes income, or even whether it redistributes from the rich to the poor or from the poor to the rich. A simple flat tax has been proposed to ensure that all people pay an equal percentage of their income in taxes. However, this would do nothing to address the fundamental issue of economic equity and justice.

A far more equitable and just proposal might be to institute a "marginal" flat tax, first granting everyone an equal "tax credit," and then ensuring that all are taxed an equal percentage on whatever income they earn. Such a tax would work similar in concept to a "negative income tax" proposed by the economic conservative, Milton Friedman.[3] The tax credit could be set at a level sufficient to ensure economic survival of all. Each dollar of income earned would then result in a net addition to disposable income, providing an incentive for everyone to work and earn. The tax credit would supplement earned income, in essence being a negative income tax, until the taxpayer earned enough to owe taxes greater than the amount of the credit. Beyond that point, the taxpayer would owe taxes on amounts greater than the credit, which would be used to fund the supplements to incomes of those with lower incomes. Such a tax credit would be insignificant to the rich, but might be critical to the survival of the poor. A flat "marginal" tax would seem far more equitable and just than the current tax, or a flat "average" tax.

State and local governments have roles and functions different from the federal government. Some public goods and services should be made available to everyone in the nation, some legitimately can be left to the discretion of states, and others are fundamentally local matters. Most of the fundamental rights discussed previously are rights to be shared equally by everyone in the nation. These public goods and services should be supported by federal taxes and administered by the federal government. The states should not be forced to raise state taxes to pay for national public goods and services mandated from the federal level.

However, individual states and cities might choose to provide some higher level of public service is guaranteed to all, and such state and local supplements logically should be supported by state and local taxes. Many collective purchase decisions are made at the state and local levels, and thus should be paid for from tax revenues collected at the level receiving the services. Nothing that can be done fairly and effectively at the local level should be done by the state level and nothing that can be done fairly and effectively at the state level should be done at the federal level. Again, this just makes sense.

So how is this new societal vision different from the societal situation that exists today? First, and most important, the new vision is of a society that understands and embraces the legitimate role of government, along with other public service organizations, in serving not just our collective interest, but also our common good. For the common good, relationships are of critical importance, and positive relationships among people within societies demand a sense of social equity and justice. This is the new societal vision: a vision of "equitable and just societies."

Equity and justice in America will require a new consensus on some new public goods and services. For example, the new vision is of an American consensus asserting that all people have a basic right to some minimum level of health care. Americans already have agreed we should provide health care to retired people, through Medicare, and to poor people, through Medicaid. Some states have instituted public programs to ensure adequate health care for all children, regardless of the ability of their parents to pay. However, widespread disagreements exist concerning how public health care should be provided; a single-payer program administered through government or a subsidized health insurance program administered by the private sector seem the be the two most popular alternatives. Regardless of the approach, we ultimately must concede that we simply can't afford all possible health care for all people, and decide what level of health care is truly a basic right, just as we have done for public education. The new societal vision includes an American consensus that some level of health care is a legitimate public good.

Once we have agreed that basic nutrition, education, and health care are American rights, we may then also conclude that they are human rights. We might shift some portion of defense and international military-assistance programs to international health, education, and agricultural assistance, and thereby help build a far safer, more sustainable world.

The new societal vision also includes a new national consensus concerning protection of the natural environment. Current environmental programs are based on the rights of people to be protected from the negative effects of pollution on health and quality of life. The conceptual basis for such programs is protection of one's person or private property from being damaged by another. Protecting natural resources held in common for the edification and enjoyment of the public seems mostly a matter of managing collective goods and services. Continuing political conflicts concerning the legitimacy and priority of environmental policy reflect a lack of consensus concerning our fundamental public responsibility to protect the environment for the benefit, not only for ourselves, but also of those of future generations.

Protection of natural resources for the benefit of future generations is an act of stewardship – taking care of something for the sole benefit of others. True stewardship is a fundamentally ethical, moral act. Until fairly recently, stewardship was a personal matter – a matter of individual values and personal choice. As long as humanity seemed technologically incapable of damaging the earth beyond its ability to restore and regenerate itself, anything an individual did to their immediate environment was considered a personal matter. "It's my land and I can do whatever I please with it." If they infringed on the private property rights of their neighbors, that was a matter to be settled between the two of them, perhaps in court, but nonetheless, still a personal matter.

It is obvious to most of us that humanity now possesses technologies capable of seriously degrading, if not destroying, the planet – or at least making it uninhabitable for humans. Protection of the biosphere so that it will continue to sustain human life on earth has become an issue of common interest because it involves the common good. Individual acts of stewardship are no longer sufficient to ensure long run sustainability of life on the planet, because the economic incentives for exploitation are simply far too strong. Corporations have no sense of ethics or morality, thus they have no incentive for true stewardship, and corporations are gaining control of more and more of the earth's resources. They are incapable of true stewardship and thus must be made to conform to the environmental consensus of society through government.

Eventually, there must be a national consensus concerning the rights of future generations to a clean and productive environment. This will represent a major philosophic step forward for humanity. A constitutional amendment to protect the environment would reflect a national consensus that the people of future generations, as well as those of the present, have fundamental civil rights. This would represent a major milestone in our transition to a more sustainable human society. And we will need governments of the people and by the people to establish and implement this fundamental principle of sustainability.

Agricultural policy today provides a prime example of what's wrong with U.S. public policy in general. Farm policy is no longer about stabilizing agricultural production and food prices; instead, it is about subsidizing wealthy landowners and agribusiness corporations with taxpayer dollars. Most government payments today are direct payments to owners of farmland who can produce as much as they choose or produce nothing at all, and still receive their government payments. Government subsidies ensure chronic surpluses of basic agricultural commodities, primarily benefiting the agribusiness corporations that process, transport, and export those commodities, but do nothing to ensure stable prices in the supermarkets at home. Government payments allow current landowners to acquire still more land, resulting in ever-larger and ever-fewer farms and farm families, which results in the continual decline and decay of rural communities. In addition, the large specialized, industrialized farming operations, made possible by government programs, pollute the rural environment with agricultural chemicals and animal wastes, while eroding and degrading the productivity of the land, which threatens the long run food security of the nation. The agricultural lobbyists for farm programs today are committed to getting as many taxpayer dollars as possible for their particular constituency, not to serving the common good through agriculture. And the most powerful among these lobbyists are the giant agribusiness corporations.

Among the most important responsibilities of government is public oversight of the private sector. The government is responsible for making sure that the corporate sector of the private economy works for the public good. The government is the only means available to restore the competitiveness of markets and thereby restoring capitalism to the American economy. Once again, as in the early 1900s, the government must break up the large corporations, and restore a competitive market structure. When a market has a sufficient number of small, independent buyers and sellers so that no single buyer or seller can have an effect on market price or output, price fixing, marketing sharing, and other non-competitive market conduct becomes far more difficult. If corporations have market power, they eventually will use it. The government is responsible for ensuring that they don't have it, and if they have it, to take it away. We did it during the Progressive era of the early 1900s and we must do it again.

With the corporations broken, and their political power gone, billions of tax dollars currently going for corporate tax credits, public-private partnerships, and other forms of corporate welfare will be freed for use in truly serving the common good. After eliminating all corporate welfare, the new societal vision of a more proactive government will quite likely cost the average taxpayer even less than we are paying today, while providing far more truly public benefits.

Yes, it is our money and we should decide how to spend it. But, we are not wasting money when we spend it to ensure the inalienable rights of all people. We are not wasting money when we spend it to ensure the integrity of our democratic society and our capitalistic economy. And we are not even wasting money when we decide to buy things collectively rather than individually. Instead, we are helping to build a more equitable and just human society. This is the new societal vision for the future.

The future of humanity depends of our willingness and ability to work together for the common good. Nothing can prevent us from restoring the integrity of our government, if we choose to do so. Nothing can prevent us from reforming our government so that it functions for the common good, if we chose to do so. And nothing can prevent us from building a more equitable and just future for humanity, if we choose to do so.

Endnotes

[1] *Growing a Nation, a History of American Agriculture*, Economic Research Service, U.S. Department of Agriculture, <http://www.agclassroom.org/gan/timeline/farmers_land.htm> (accessed September 2006).

[2] Robert D. Putnam, *Bowling Alone* (New York: Simon and Schuster, Inc. 2000).

[3] Milton Friedman, *Capitalism and Freedom* (Chicago: University of Chicago Press, 2002, original copyright, 1962).

10

AN ECONOMICS OF SUSTAINABILITY:
A NEW ECONOMIC VISION

In the summer of 1996, I was invited to present a paper at a conference on sustainable agriculture sponsored by the American Agricultural Economics Association meetings in San Antonio, Texas. The conference organizers assigned me the topic of, "Sustaining the Profitability of Agriculture." In thinking about how I should approach the subject, I concluded that the economy is not designed to sustain the profitability of agriculture. In fact, our economy virtually ensured that any profits in agriculture are quickly competed away, and thus farm profits are inherently unsustainable. The same would be true of any economically competitive sector of the economy. Any expectation of profits would invariable lead to increased production, which would cause prices to fall until profits disappear. The reason corporations are trying to gain control of agricultural markets is so they can eliminate competition, restrict total production, and sustain profits through their power to influence the marketplace. But for farmers, who have no market power, any profits that happen to arise quite simply are not sustainable.

At the conference, I urged agricultural economists to accept the challenge of developing a new theory of economics, a theory that would allow reasonable levels of profits to be sustained, even with economically competitive markets. I was suggesting that economics should accept responsibility for the social and ethical implications of a free market economy. The discipline of economics simply doesn't give much attention to what the inherent lack of sustainability of profits does to our overall quality of life. In economics, profits provide the economic incentives and the financial means for continual innovation, so that we can have ever more stuff at an ever-lower price. But the personal fate of those who don't survive the race to produce more and produce cheaper doesn't get much attention. Some win and some lose; that's just accepted as the discipline of the marketplace. Neither does the "economic man" spend many sleepless nights worrying about what this blind pursuit of materialism does to human society in general. Consumer satisfaction is all that matters in economics. The social and ethical well-being of people as members of society must be addressed outside of economics.

At the meeting in San Antonio, I argued that a new economics should be developed upon the foundation of sustainability – upon ecological integrity, social responsible, and economic viability. I challenged my colleagues to break out of their myopic ways of thinking and to examine their blind faith that short-run self-interests somehow will be transformed into long-run societal interests. I challenged them to abandon the dogma that we had been taught in graduate school and to develop a new economics of sustainability.

At lunch, we had a table of people who seemed ready to join my campaign for a new kind of economics. A former colleague from Oklahoma and I had a long conversation following lunch that day. He had spent his career in resource economics – trying to find ways to minimize the negative impacts of the private economy on the natural environment. He agreed with most of what I had to say, but he didn't think we needed to develop a "new economics" – "the old one was not beyond repair," he said. I disagreed. He challenged me to take the first step, to propose a new economics of sustainability – to outline something that wasn't just a rehash of the old economics. I accepted the challenge.

Although I have dealt with some aspects of this new economics in earlier chapters, this chapter brings them together to form a new economic vision for the future. The concept of sustainability is far broader than economics – at least economics as it's currently conceived. Herman Daly and John Cobb, in their book, *For the Common Good*, refer to the economics of today as "chrematistics" – the "manipulation of property and wealth so as to maximize short-term monetary exchange value to the owner."[1] Sustainability is also broader than ecology or sociology, because sustainability most certainly includes economics. However, sustainability is quite consistent with the Greek root-word for economics, "oikonomia" – meaning administration or management of the household so as to increase its value to all members over the long run. The concept is equally applicable to management of the community, society, humanity, or the biosphere. Thus economics, as oikonomia, includes the management aspects of sociology and ecology as well as economics.

Daly and Cobb propose to address oikonomia through an economics of community, which they propose to achieve through new government policies. New public policies will be necessary, but it will take more than new public policy to implement oikonomia in America. First, the people must embrace this new concept of economics as a part of American culture. They must understand and appreciate the necessity for managing or administering our society, environment, and economy as a unified, inseparable whole.

Sustainability is a long run, people-centered concept. The purpose of sustainable development is to sustain a desirable quality of life for people over the long run, forever. Granted, with our limited knowledge, we cannot conceive of being able to sustain life on earth without a continuing inflow of solar energy. Thus solar-powered systems of production represent the current limit to our thinking with respect to means of achieving sustainability. Perhaps in some distant post-solar era, life will be spiritual rather than physical in nature and will not require energy. However, lacking any real understanding of what might happen once the sun burns out, and in the absence of any logical endpoint of time, the purpose of sustainability remains: to sustain a desirable quality of human life on earth forever.

Some find fault with this human-centered or anthropocentric approach to sustainability. Some deep ecologists, for example, contend that other forms of life may be just as important as human life in the longer run scheme of things. However, if we are not concerned, uniquely, with sustaining progress of the human species, there is no economic issue to be addressed. Economics is about managing resources to meet the needs of humans. If we weren't particularly concerned about humans, we could simply depopulate the earth, or otherwise reduce human claims on the earth's resources, to a point where the sustainability of other species would no longer be in question, or at least not threatened by humans.

However, our nature as humans is not unlike the nature of other species, in that we humans have an innate instinct for survival, reproduction, and self-gratification. We will not reduce our claims on earth's resources for the sole purpose of ensuring the sustainability of other species or of the earth. However, we will protect other species if we perceive it to be in our best interest to do so. The fact that we are concerned uniquely with sustaining the human species, does not imply that we are concerned exclusively with sustaining the human species. Contrary to what the economics of chrematistics might imply, our quality of life is not exclusively individualistic in nature. Our interests as members of our particular societies and as members of the human race are integrally linked with the integrity of the rest of the biosphere. Thus, our interests as human beings may well be served best through sharing and stewardship, including preservation of other species, rather than through expressing our individual human greed.

As I initially conceived the new sustainable economy, it was an inseparable whole with three different dimensions – the individual or private economy, the social or public economy, and the ecological or moral economy. The three are clearly interconnected and interdependent, but each has distinct characteristics that warrant their separate consideration.

The ecological economy provides the foundation for the other two. People exist upon the face of the earth and cannot exist without the resources of the earth, although many have lost sight of this fact in the highly specialized, industrial economy of today. Humans cannot live without sunlight, air, water, and soil. Most have been reminded by industrial pollution of our air and water of our tenuous dependence on the resources of the earth. We became concerned about air and water only when supplies of clean air and water became scarce. However, many never stop to think that we are equally dependent upon the soil. Pure air and pure water will not support life. Life requires minerals from the earth in addition to air, water, and sunlight. All of life ultimately arises from the soil – even life that resides in the ocean. Without the mineral and biological resources of the earth, life on earth is not possible.

Life ultimately depends on energy. Humans get the energy to fuel their bodies from other living things, including both plants and animals. We can't get enough energy directly from the sun or the earth to support human life. Living organisms, mostly plants, convert solar energy, soil, water, and air into energy forms that can be consumed by animals, including us humans. All energy present on the earth today was either stored in the earth at its time of creation or has been captured from the sun and stored since then by living organisms. Fossil fuels are nothing more than energy captured and stored by prehistoric plants and animals. As indicated in Chapter 8, energy is neither destroyed nor created on earth, but each time energy is used and reused, some of its usefulness is lost. The only source of new energy available to offset the inevitable loss of usefulness of energy to entropy is solar energy from the sun.

We humans currently are using fossil energy far faster than we are capturing new solar energy from the sun. We are using up the resources of the earth and eventually will run out, if we don't bring our rate of depletion into balance with the rate of renewal. We ultimately must balance the loss of useful energy with new energy captured from the sun, or the earth will inevitably trend toward an absence of form, pattern, hierarchy, or differentiation – toward entropy.

We also must use the earth as a dump or a "sink" for our wastes, as well as a source of energy to sustain life. Wastes represent depleted energy, because we don't know how to use it, and pollution represents negative energy, because it takes energy to mitigate or neutralize it. There is no place else to dump our wastes except into the natural environment.

Compost of lives

deplete environment *erosion*

When we dump waste into the environment, we are degrading or destroying its usefulness to humans – as when we pollute air or water, or when we poison the soil with chemicals or allow the soil to erode away. Some things that we dump into the environment could be reused or recycled, turning wastes into useful resources. However, some waste is unavoidable because of entropy. So, we lose the use of ecological resources, both through avoidable waste and unavoidable entropy.

As far as we know, the first and second laws of thermodynamics cannot be repealed. We cannot continue indefinitely to use up the resources of the earth at rates faster than their rate of regeneration. A sustainable ecological economy is one where our use of energy and matter, as resources and as sinks, is equal to or less than the inflow of solar energy and the energy extracted and reclaimed from the natural environment. The laws of nature cannot be changed. Ultimately, we must reconcile the individual and social economies with the ecological economy, if we are to sustain life, including human life, on earth.

The environmental movement has made the public aware of this ecological economy. However, the concept of a social economy is neither well understood nor appreciated. First, as indicated in Chapter 4, the social economy is not the macro-economy – the aggregate of individual enterprises. The social economy addresses the interconnectedness of people within society – not just the adding together of individuals. The purpose of the social economy is to facilitate the building of a stronger society by building and sustaining positive relationships among people, as indicated in Chapter 9.

Social capital is the essence of any civilized society. Social capital includes the ability of people to relate to each other, to form families, communities, and nations, to agree on processes of governance and trade, and to share basic principles and values by which civilized people agree to live.[2] The indicators of social disconnectedness mentioned in Chapter 9 measure losses of social capital, reflecting the loss of caring and civility in relationships. Analogous to stocks of fossil energy, our current stocks of social capital have been built up over centuries by past human civilizations. Without social resources or social capital, we would be living in barbaric anarchy. Our social resources are at least as important as our ecological resources in supporting our current quality of life.

Stocks of social capital are built up through human culture – the passing from generation to generation of the lessons learned through the struggles of people to achieve more peaceful, productive, and harmonious relationships. Anything that helps us to achieve a higher quality of life through human relationships builds the stock of social resources. Wars, feuds, confrontations, arguments, destructive competition, and other forms of human conflict destroy social capital. People who have lived in peace for centuries can become bitter enemies as a consequence of simple misunderstandings. Misunderstandings can lead to conflicts, conflicts to confrontations, and confrontation to wars. The ability to live and work together is destroyed, the social capital is depleted, and quality of life is diminished.

Stocks of social capital are inevitably depleted through ordinary acts of human incivility. This unavoidable loss of social capital is analogous to entropy in the ecological world. Thus, social capital must be continually replenished or restocked if human relationships are to remain positive and the quality of life within a society is to be enhanced over time. Social capital can be depleted also through deliberate acts of oppression, exploitation, discrimination, injustice, or even indifference. Such acts became commonplace in the latter stages of American capitalism and have become rampant in our corporatist society of today. Corporatism places no value on human relationships, other than those that can be exploited for economic gain. Corporations have no family, no community, no nationality, nor a sense of belonging. Corporations destroy the social fabric of families, communities, and nations in their pursuit of profits and growth.

If human civilization is to continue to advance in the future, we must maintain and continually build our stocks of social resources. Thus, the new sustainable economy must be designed to halt the senseless depletion of social capital by corporate exploitation. But the new economy must also encourage and support continual social investments by encouraging positive human relationships. Stocks of social capital must be built at rates exceeding their natural rates of erosion plus the rates of unnecessary depletion, if human civilization is to be advanced. This is the fundamental nature of the social economy. The progress of human society demands that our addition to stocks of social capital consistently exceed our continuing withdrawals.

The conventional or individual economy is by far the best understood and most widely appreciated of the three economies. It's the only concept most people associate with the word economy. When we go to work or go shopping, we are participating in the individual or private economy. The private economy provides the means by which we meet our needs as individuals, and as collections of individuals, through our transactions with other people and our interactions with the natural environment. As indicated before, if we lived totally independent and self-sufficient lives, we would have no need for an economy. But our lives can be made better through specialization and trade, and thus, we need to relate to other people. Our lives can be made better through utilization of natural resources that are beyond our physical grasp, thus we need to trade to acquire benefits from those resources to which we would not otherwise have access.

The private economy doesn't use ecological and social resources directly, but instead converts them into economic resources. Ecological resources are extracted from the natural environment – through mining, logging, or farming – and are converted into marketable raw materials for manufacturing, construction, or processing. Natural resources that were once owned in common, for the benefits to all, are converted into private goods for sale to the highest bidder. Without privatization, however, there would be no incentive for private investment in the resource extraction process, and nothing would be produced for sale.

Economic resources are extracted from society – through employment, collaboration, or negotiation – and are converted into economically valuable human resources that produce goods, provide services, or make deals. Social resources that once supported positive relationships among people are converted in labor, joint ventures, and commercial advantages – commodities for sale to the highest bidder. Without privatization of these social resources, however, there would be no individual incentive to make deals and or invest in enterprises that employ people, and there would be no jobs in the private sector.

The macro-economy is a part of the individual economy. Macroeconomic policy plays a legitimate and important role in the function of the individual economy, however, it does nothing to sustain or support the social or ecological economics. In fact, as indicated in Chapter 4, current macroeconomic policies, in promoting maximum economic growth, promote the degradation of both social and ecological resources.

The individual or private economy is an important dimension of any modern society. The conversion of ecological and social resources into economic resources is both necessary and legitimate, if we are to live at any level above subsistent self-sufficiency and have the freedom of individual choice. However, we must recognize that when we convert natural resources into economic resources, fewer natural resources are left to support the ecological economy. When we take minerals from the earth, cut old-growth forests, or farm ecologically fragile soil, we are disrupting the natural ecosystem in ways that may degrade its ability to remain healthy and productive.

We obviously benefit from natural resources indirectly through the economy, but we must also recognize that we benefit directly from the natural environment – from breathing fresh air, drinking pure water, and from being good stewards of the air, water, and soil. Whenever we use the private economy to extract from nature at rates faster than nature can regenerate; we are degrading the long run productivity of the ecological economy, and ultimately will degrade our overall quality of life.

We must recognize also that when we convert social resources into economic resources, fewer social resources are left to support the social economy. When we go from helping each other voluntarily to working for each other, we have transformed a personal relationship into a business arrangement. When we start using our social contacts with other people as business connections, we have started to transform friends into business prospects and social gatherings into business conferences. Ultimately, we will begin to compete rather than cooperate. The means by which we relate to each other will become defined by common business practices, rules, or laws rather than by a sense of caring and common human understanding. Certainly, humans benefit indirectly when friendships spawn business relationships, but we must realize also that humans also benefit directly from our purely social relationships – from belonging, caring, sharing, and loving. When we use the conventional economy to extract social resources at rates faster than the rates at which we reinvest in society, we degrade the productivity of the social economy, and ultimately will degrade our quality of life.

An economics of sustainability must integrate the individual, social, and ecological economies. The purpose of sustainable economic policy must be to promote harmony and balance among the individual, social, and ecological economies. The three economies are inseparable dimensions of the same whole. We simply cannot do anything that affects one of the three economies without affecting the other two. If we are at a point of balance or harmony, we can't do anything to improve one dimension without disturbing the balance, and thus, making ourselves worse off overall than before. If we are out of balance, we can restore balance by devoting less time and energy to the dimension that we are overdoing, and improve our overall well-being by actually doing less. An economics of sustainability will force us to rethink the conventional wisdom of maximizing income and wealth, and instead seek harmony and balance as a means to a more desirable quality of life.

There is always an element of tension or stress among the economic, social, and ecological dimensions of our lives – if for no other reasons than all demand investments of our limited time and energy. However, tension is not the same as conflict, and stress is not the same as distress. Tension can be positive in that it is often necessary to build strength. Similarly, there always will be tensions and stress in a sustainable economy, even when the individual, social, and ecological economies are in harmony. Tensions become conflict and stress becomes distress only when one dimension is made stronger by weakening the others. With positive stress, the strengthening of one dimension creates a healthy tension that challenges, encourages, and eventually strengthens the other dimensions as well.

Admittedly, the most difficult challenges in developing a sustainable economy are likely to arise from the integration of its economic, social, and ecological dimensions – in maintaining a positive, dynamic balance or harmony among the three. This will require an effective means of resolving short run conflicts or contradictions. Some issues clearly relate to the individual economy – the costs and benefits accrue almost exclusively to individuals. Other issues are clearly social – individuals must work together through government to ensure equity and justice. In fact, most short run environmental issues are addressed as social issues, by ensuring the rights of all people within society to be protected from the health and environmental risks associated with pollution.

Longrun ecological and social issues are fundamentally matters of morality – this generation accepts the responsibility to protect the rights of future generations as a matter of ethical or moral principle. Future generations cannot participate in today's marketplace, to buy resources for their future use, and neither can they vote in today's elections to ensure that resources are set aside by the government for their use. They must depend upon the moral and ethical sense of stewardship among those of today's generation. Those of today's generation will receive no economic or social benefit from their decisions to protect or set aside natural resources or to preserve the civility of society for the benefit of future generations. Those of current generations will not be around seven or seventy generations in the future to receive any economic benefits and do not have social relationships with anyone who will.

The challenges arise from issues that have important economic, social, and ecological dimensions – at the margins or intersections among the three. Sustainable systems are not hierarchical in the conventional sense of one organizational level dominating another. However, they are hierarchical in a systemic sense, in that the economy is a subsystem of society, which in turn is a subsystem of the natural ecosystem. Nature might seem to be dominant over society and society dominant over the economy. However, the economy can create consequences that either enhance or destroy society and society can make choices that either enhance or destroy natural ecosystems. So an interdependent relationship exists among the three – none can survive independent of the other. Of course, nature might well survive the ravages of both the economy and our current society, but it likely would be a nature incapable of sustaining contemporary human society.

The hierarchy of sustainability arises from the source of organizational principles or rules by which the system as a whole must function. The concept of ecology presumes there are inviolate laws of nature – a higher order of things within which all else, including human society, ultimately must find harmony. The economy is a creation of society, and thus society sets the rules by which the economy must function. Violation of this hierarchy principle is neither impossible nor uncommon, but continual or egregious violations, quite simply, are not sustainable.

The natural hierarchy among ecological, social, and economic systems suggests a necessary hierarchy of human rights and responsibilities. The rights of humanity first must reflect our ethical and moral responsibilities to human civilization – not to degrade or destroy the ecological and social resources upon which the future of humanity must depend. Only within this context of basic human rights, can we fully realize our rights as citizens, to be a part of an equitable, just, and productive society. Our economic rights, to be protected from economic exploitation and rewarded in relation to our productivity, can be sustained only within the context of a civil society. Thus, issues of conflict can be resolved, conceptually, by relying on this natural hierarchy. The challenge of translating this concept into reality, which will be addressed in Chapter 14, may prove somewhat more difficult.

Since that meeting in 1996, I have been working to refine my initial thoughts on the economics of sustainability. I have even written a book, *Sustainable Capitalism: A Matter of Common Sense*.[3] However, I certainly do not claim to have developed the complete blueprint for a new discipline of sustainable economics. In fact, I have barely scratched the surface of what ultimately needs to be done. However, a new economics of sustainability can be developed, and hopefully will be developed within the next few decades. The necessary components of this new economics already exist. We only need to put them together to form a new coherent whole.

We already have a private economy that could be fixed to pursue our individual interests. We have a democratic government that could be used to pursue our social interests. And, we have a Constitution that could be amended to reflect more fully our moral and ethical values. All we need now is a shared vision of how the individual economy, the social economy, and the ecological economy should work together to support and sustain a more desirable quality of life. We need a vision of economics as oikonomia (managing, for long run benefit the whole) rather than chrematistics (manipulating, for the short run benefit the individual). With this shared vision to guide them, people can begin to create the new parts that will be necessary to create a new whole, a new economics of sustainability, an economics as if people really mattered.

How does the new economics answer the initial question of how farmers might expect to sustain profitability in a sustainable society? First, in a sustainable economy, neither farmers nor any other producers would attempt to maximize profits, but instead would attempt to find balance among the economic, social, and ecological dimensions of their operations. They would be pursuing a more enlightened concept of self-interests. So the profits of producers that are less efficient from a strictly economic perspective would not necessarily be eliminated by competition from more economically efficient producers. Stewardship of the environment and contribution to the community would be recognized as attributes of total productivity. Farmers would not feel compelled to drive their neighbors out of business, if they fully considered the negative social impacts of fewer farmers in their community and the negative ecological impacts of having fewer farmers to care for the land.

Second, public policies in a sustainable society would not allow people to be driven out of business, if their contribution to the social and ecological well-being of their community and society more than offset any loss of economic efficiency. Their neighbors, and society as a whole, would recognize their positive contribution to the overall societal well-being and would create the economic environment necessary to ensure their continued contribution. Our economic policies today are designed to minimize costs by driving less efficient producers out of business, with little regard for the ecological and social consequences. Sustainable economic policies would ensure the economic sustainability of ecologically sound and socially responsible business operations.

Over time, as more and more consumers recognize and become willing to pay the full economic, ecological, and social costs of food production, farmers who protect the environment and contribute to society will realize their profits from the marketplace. And the profits will be sustainable, because they will reflect the unique contributions of individuals who are committed to sustaining the productivity of their unique natural and social resources – their land and their communities. The profits of those who fail to make a positive contribution to overall societal well-being would not be sustainable. If they refuse to change, they would be allowed to fail. But profits of those who choose to help ensure the sustainability of humanity would be sustainable.

Decisions in the sustainable economy should be based on the "true value" of things – not just on economic value. But what is the "true value" of things? This has been a nagging question in the back of my mind ever since my graduate school days. I used to have long discussions with my fellow graduate students at the University of Missouri about market value, intrinsic value, consumer surplus, and welfare economics, all of which attempt to determine the value of things. However, I was never satisfied with the outcome. I always argued that the true value of things was different from any of the measures of value that we discussed in economics, but I didn't know how to explain what I meant. I believe that an economics of sustainability, ultimately, will have to address the concept of true value.

First, I have come to believe that there is no single true value of anything. The economic value of anything is a value determined by its degree of scarcity in a competitive marketplace, as explained in Chapter 2. Each person votes with whatever dollars they have, or can get, and the price paid by the highest bidder determines the economic value. Economists talk about different concepts of economic value. Things have average values, marginal values, total values, and all-or-nothing values, to name a few. But, these are all economic values, determined by scarcity – the relationship between supply and demand – and all can be measured in terms of dollars and cents. Economics provide but one means of determining value.

The social value of a thing is different from its economic value, although economists mistakenly attempt to assess social value by using economic measures. In a democracy, the social value of anything must be determined by giving every participant an equal voice in the valuation process. In a democratic society, we are all of equal inherent social worth. None is considered to be wise or good enough to make our societal choices for us. If we wanted to discover the social value of something in the marketplace, each person would have to be given the same number of dollars to spend so they would have an equal voice in determining the value of whatever was offered for sale. In the real world, the nearest we come to measuring social value is through the process of buying and selling public goods and services. In these cases, the government acts on behalf of the people, and theoretically, each person has an equal voice in influencing their public purchase decisions. In social evaluation, each person must be given an equal voice. Thus, we should never expect public decisions to buy or not buy things to reflect economic values. Economic and social values are fundamentally different.

In addition, the ethical or moral value of anything is different from its economic or social value. The moral value of a thing cannot be determined in either the marketplace or the voting booth, but must be determined by a process of consensus. The moral or ethical value of anything must reflect our common sense of its worth. To determine the moral value of a thing we would have to bring people together, without money and without votes, with a commitment to agree on the relative worth of things. They would have to come to a consensus. In all probability, they would not rank the value of diamonds as high as the value of air, nor rank the value of a sports hero or rock star as high as the value of a school-teacher or a nurse. I have no doubt that such a group eventually could reach a consensus regarding the ethical or moral value of things. And such a process ultimately must be used to determine long-run values of things in the ecological and social economies. Long run ecological and social values of things cannot be measured in dollars and cents nor determined by votes, but must reflect a moral consensus regarding their worth.

The "true value" of a thing thus must somehow reflect the whole of its economic, social, and ecological value to humanity. More important, however, is the fact that there is no single best measure of true value, in spite of the common practice of giving economic value priority over the other two. In addition, there is no means of converting one value into another to derive a single measure of true value, in spite to the popular practice of converting social and ecological values into economic terms. Finally, the maximum true value of a thing will be realized when there is balance and harmony among its economic, social, and moral values. The search for an economics of sustainability should bring us closer to real-izing the maximum true value from the individual, social, and ecological dimensions of the sustainable economy.

In 1998, at a small bed and breakfast place upstream from Koskie, Idaho, I met with a small group of economists and others scientists from around the country who were interested in pursuing the idea of devel-oping a sustainable economy. We concluded that the new economics must begin with a Declaration of Interdependence, a statement that a sustainable economy must recognize the fact that all things, living and non-living, are interconnected and interdependent, and that the sustain-ability of the parts, including the human race, depends upon the sustain-ability of the whole. This recognition is a direct contradiction of conven-tional neoclassical economic thinking, which views society as a collec-tion of individuals and society and the environment as external con-straints to economic progress. We concluded that we were not likely to create a sustainable economy unless at least a small group of thoughtful individuals was willing to commit their lives and their fortunes to such a cause, as did the framers of the Declaration of Independence in 1776.

In this chapter, I have attempted to present my vision for a new sus-tainable economy – an economy that will facilitate the pursuit of a more enlightened self-interest, by managing the individual, social, and ecologi-cal economies for long run sustainability. I believe this is a vision worthy of the commitment of the lives and fortunes of those of us who share it. And it is built upon a foundation of common sense.

Endnotes

[1] Herman Daly and John Cobb, *For the Common Good: Redirecting the Economy Toward Community, the Environment and Sustainable Future* (Boston, MA: Beacon Press, 1989).

[2] James S. Coleman, "Social Capital in the Creation of Human Capital," *American Journal of Sociology Supplement*, 94, (1988), 95-120.

[3] John Ikerd, *Sustainable Capitalism: A Matter of Common Sense* (Bloomfield, CT: Kumarian Press, Inc., 2005).

11

THE GREAT TRANSFORMATION:
A VISION OF CHANGE

In nature, every living thing eventually must die, so that new things can be born, live, grow, mature, and produce new life. Likewise, the industrial era of development eventually must die so that a new era of sustainable development can be born, grow, mature, and renew. Trends, such as industrialization, never continue forever. Each era of human history passes on, to be replaced by another and then another. The age of enlightenment replaced the dark ages and the industrial era replaced an era of crafts and guilds. Today, human civilization is again passing through a time of great transformation. An old age is dying and a new age is being born. Human progress is moving onward to a new and hopefully better future.

A few years back, a couple of young scientists proposed a list of the top twenty "great ideas in science."[1] The proposal was reported in the journal *Science*, one of the two most respected scientific journals in the world. The authors invited scientists from around the world to comment on their proposed list. Among the top twenty were such ideas as the laws of gravity and motion and the first and second laws of thermodynamics. The top twenty also included the proposition, "everything on the earth operates in cycles" - everything, physical, social, biological, and so on. The planets revolve around the sun, the seasons come and go; people are born, live, reproduce, and die; political ideology swings from liberal to conservative and back again; the economy goes from boom to bust, federal budgets from deficits to surplus; women's hemlines go up and down and men's ties get wide and then narrow again. Everything on earth operates in cycles. Some scientists responding to the article proposed a slight revision to the theory of universal cycles, suggesting that everything "tends" to operate in cycles.[2] But the theory of universal cycles was confirmed to belong on the list of the "top twenty great ideas in science."

In essence, the theory claims that no trend continues forever. Trends are nothing more than phases on longer-term cycles. Industrialization was simply a trend that replaced the previous trend of crafts and guilds, which had run - actually had overrun - its useful course. Paradigms, like industrialization, become popular because they seem to solve many of the pressing problems of the times. Eventually, most of the problems solvable with the new paradigm are solved, but in the process, people become convinced that the new paradigm will provide a solution for all problems. As a result, it is used in situations where it has no value, and eventually it begins to be used in situations where it creates more problems than it solves. A new paradigm then must be developed to solve the problems created by misuse of the old paradigm, and to solve those problems the old paradigm was inherently incapable of solving. Nothing was necessarily wrong with the old paradigm; it's just that all paradigms have limitations. Regardless, the overuse and misuse of old paradigms inevitably leads to their rejection and to the acceptance of new and different paradigms to replace them. We are now in midst of such a transformation.

The industrial era has now run its course. The corporatist phase has clearly taken the industrial process beyond usefulness to destructiveness. The time has come for a new paradigm for sustainable development and a new era of sustainable ecological, social, and economic progress. The new paradigm of sustainability will solve the problems created by industrialization and address a host of new opportunities for the future. No matter how improbable it might seem to some, the corporate industrial era is ending; the question is not if, but when and how.

Gail Imig was a Vice President and the Director of University Extension during the early 1990s, just after I returned to Missouri. She was one of the few university administrators I met who was truly able to think outside of the "academic box." Lots of them talked about it, but few actually did it. She was one of the few people at the University at the time who understood that human society was going through a great transformation - so fundamental that it ultimately will affect every aspect of our lives.

I became acquainted with Gail while I was working on the proposal for the sustainable community development project, mentioned in Chapter 8. One day in Gail's office, she asked if I was familiar with the writings of Alvin Toffler, the futurist. I told her that I had heard the name and had read a magazine article by him, but had never read any of his books. She said Toffler had been writing about things that seemed very similar to what I was talking about, and gave me a copy of a recent article to read. The article referred to some of the things he had written in his book, *Power Shift*.[3] I became excited as I read the article, so I bought the book. Here was a highly respected author and thinker, an advisor to people as well known and diverse as Bill Clinton and Newt Gingrich – with a vision for a new society beyond the industrial era.

Reading Toffler led me to other futurists, many of whom were writing about quite similar visions for the future. These writers were selling millions of books and reaching millions of people, and I now sensed the number of people thinking about similar things weren't just a few, but were many. I hadn't learned what I believed about the great transformation from them and they certainly hadn't learned what they had written from me. Yet, we were thinking and writing essentially the same things. Modern human society is going through a great transformation, out of the old industrial era and into something fundamentally new and different. Now I knew I wasn't just daydreaming; what I had seen was real.

I have purposely minimized the use of citations in this book, so far. If what I write doesn't make sense on its own, I don't expect anyone to believe it, regardless of how well referenced it might be. And if something makes sense, I know people will believe it, not because it is well documented, but because their intellectual insight – their common sense – tells them, it is true. I don't believe that anything is necessarily more valid just because someone else has written it down somewhere, nor is it any less valid just because this is the first time it has been written. No idea is original. Everything any of us knows today is made up of ideas that have been floating around the universe since the beginning of time. The only concept of "new" that we know is "new arrangements" of old ideas to create new wholes.

That said, in some situations, citations are useful. In reading Robert Pirsig's, *Zen and the Art of Motorcycle Maintenance*, I began to understand that the only concept we know of "objective reality" is our common understanding of reality, an understanding that is shared among thinking beings.[4] He wrote, "What guarantees the objectivity of the world in which we live is that this world is common to us with other thinking beings. Through the communications that we have with other men we receive from them ready-made harmonious reasoning. We know that these reasonings did not come from us and at the same time, we recognize in them, because of their harmony, the work of reasonable beings like ourselves. It is this harmony, this quality if you will, that is the sole basis for the only reality we can ever know."[5]

When we communicate with other thinking people, we sometimes find that they have reached the same conclusions as we, although we have reached our conclusions independently. They didn't learn what they know from us, and we didn't learn what we know from them, nor did we learn what we know from some common third source. So, we must have discovered the same reality. Since it is highly unlikely that we would have shared a common dream, what we both have seen then must be real. As Pirsig concluded, this harmony of thought among thinking beings is the only reality we can ever know – it is our common sense of what is real.

So I have concluded, it may useful at times to bring together a collection of similar thoughts from a variety of different people, as Pirsig suggests, to show patterns of "harmonious reasoning" among thinking people. We know that their reasoning did not all come from each other, so we can conclude that these thinking beings have discovered the same reality.

During the late 1980s, a number of intelligent, thinking, reasoning "futurists" were writing about a future world fundamentally different from the world of the past two-hundred years. In his book, *Power Shift*, Toffler pointed out that many forecasters simply present unrelated trends, as if they would continue indefinitely, without providing any insight into how the trends are interconnected or of outside forces likely to reverse them. Toffler concluded that the forces of industrialization had run their course and were now reversing – that the industrial era was over. The industrial models of economic progress are becoming increasingly obsolete, and industrial measures of efficiency and productivity, such as quantity and price, are no longer sufficient, he claims. Customized goods and services targeted to niche markets, continuous innovation, and value-added products; these are the trends of the future. The most important new productive resource has become knowledge. By relying more on knowledge, conventional factors of production, including land, labor, raw materials, and capital, become less important, and thus, less limiting.

Sequential, assembly-line production systems are being replaced with simultaneous systems, where individuals or small teams transform raw materials into final products. Synergism, creating wholes that are greater than the sum of their parts, is replacing specialization as the source of production efficiency. Creating value, by meeting the unique wants and needs of unique customers, is replacing low prices as the primary source of economic progress. And, synthesis is replacing analysis as the means of exploration and discovery.

Peter Senge, in his book, *The Fifth Discipline*, deals with the concept of synergy in the context of business organization.[6] Production systems of the future will embody enormous complexity with simultaneous and dynamic linkages among a multitude of interrelated factors. Humans can deal consciously with only a very small number of different things at the same time. Yet, humans are able to perform enormously complex tasks quite easily – such as driving a car in heavy traffic, playing a tennis match, or carrying on a conversation – things that still baffle the most sophisticated computerized robots. People are capable of performing such tasks routinely by using their well-developed subconscious minds. Computers and robots don't have subconscious minds.

Our subconscious mind can solve problems without our even "thinking about them." A problem that seemed to be unsolvable the night before has all sorts of alternative solutions to explore when we arise the next morning. Insights and intuition lead us toward solutions that we would never have reached with logic and reason. Our subconscious mind is capable of dealing with complexities that are beyond our own logical comprehension. In fact, the human mind may be the only mechanism capable of dealing effectively with the type of "mind work" that will be required for sustainable development.

Peter Drucker, a time-honored business consultant and author, writes of a Post-Capitalist Society.[7] Drucker's writings were the foundation for management classes in many business schools in the '50s and '60s and he has consulted with most of the corporate giants of American industry. Drucker wrote, "Every few hundred years in Western history there occurs a sharp transformation. Within a few short decades, society rearranges itself – its worldview; its basic values; its social and political structure; its arts; its key institutions. Fifty years later, there is a 'new world.' We are currently living through just such a transformation."[8]

Drucker agrees with Toffler that the most significant development of his lifetime is the shift to the knowledge society. The future, he writes, belongs to the "knowledge worker." All developed countries that were business societies are now becoming post-business, knowledge societies. There are important, fundamental differences between knowledge work and industrial work. Industrial work is fundamentally a mechanical process whereas the basic principle of knowledge work is biological. This difference has important implications in determining the "right size" for business organizations. In a mechanical world, greater efficiency is generally associated with greater size – i.e. there are economies of scale. But in a biological world, efficiency results from fitting size to function. "It would surely be counterproductive for a cockroach to be big, and equally counterproductive for the elephant to be small," Drucker writes.[9] The intelligence of a rat and a human cannot be compared; each has found ways to thrive within an ecological context occupied by the other.

This difference in organizing principles may be critically important in determining the organizational structure and location, was well as size, of economic enterprises in the future. Structure and location will be determined by purpose and function, and other things equal, the smallest effective size will be best for information and knowledge work. And, many small information enterprises can be located virtually anywhere.

Robert Reich, former U.S. Secretary of Labor, addresses future trends in the global economy in his book, *The Work of Nations*.[10] He identifies three emerging broad categories of work corresponding to emerging competitive positions within the global economy: routine production workers, service workers, and symbolic-analysts. Production workers are the "old foot soldiers of American capitalism," he wrote, and include low- and mid-level managers – foremen, line managers, clerical supervisors – in addition to traditional blue collar workers. Production workers typically work for large industrial organizations, and live by the sweat of their brow or their ability to follow directions and carry out orders, rather than by using their minds.

Service work, like production work, can entail simple and repetitive tasks. The big difference is that service work is more "personal" and requires a "human touch." This category includes occupations such as retail sales worker, waiter, janitor, cashier, child-care worker, hairdresser, flight attendant, and security guard. Service work, like production work, requires relatively little education, and most require close supervision. Services may be provided through a diversity of organizational structures, ranging from individual providers to large franchised organizations. Unlike production work, however, individual personality can be a big plus, or minus, in performing service work.

Symbolic analysts are the "mind workers" in Reich's classification scheme. They include all the problem-solvers, problem-identifiers, and strategic-brokers. They include scientists, design engineers, public relations executives, investment bankers, doctors, nurses, lawyers, real estate developers, and consultants of all types. They also include writers and editors, musicians, production designers, teachers, and even university professors. Symbolic analysts often work alone or in small teams, which are frequently connected, but only informally and flexibly, with larger organizations. The futurists agree that the future will be dominated by symbolic-analysis, by mind work, rather than by routine production or personal service work.

John Naisbitt and Patricia Aburdene in their book, *Megatrends 2000*, call the triumph of the individual the great unifying theme at the conclusion of the twentieth century.[11] They talk about greater acceptance of individual responsibility as new technologies extend the power of individuals. Their "mind workers" are called individual entrepreneurs. Over the past few years, we have seen small-scale entrepreneurs seize multi-billion-dollar markets from large, well-heeled businesses – microcomputers firms, for example. In fact, during the latter decades of the 20th-century, small firms created far more new jobs than did the old industrial corporations.

Empowered individuals, while quite capable of working alone or in small groups, seem to seek community, the "free association of individuals." Large business organizations, government bureaucracies, labor unions, and other collectives, have provided hiding places for those who have chosen to avoid responsibility. In a community, there is no place to hide. Everyone knows who is contributing and who is not. In communities, individual differences are recognized and rewarded. The sense of community, which has been all but destroyed by industrial corporatism, may well be restored by individuals empowered with knowledge. Knowledge workers will be looking for a place to be recognized, a place to belong, and not a place to hide.

Several of these futurists talk about a new electronic heartland. They contend that this new breed of mind workers will reorganize the landscape of America. They will be linked by telephones, fax machines, Federal Express, and the Internet, forming information networks that span the globe. These mind workers are free to live almost anywhere they choose, but increasingly are deciding to live in small cities, towns, and rural areas, rather than in large cities. The industrial revolution built the great cities of Europe, America, and Japan. But, as we enter the 21st century, cities have lost much of their purpose as places for people to live.

A century ago, railroads and waterways allowed raw materials and finished goods to be moved relatively inexpensively over long distances. It was more expensive to move people. So people lived in cities, near the factories, where they transformed raw materials into finished products. Today, multi-lane freeways and extended mass transit systems have allowed people to retreat to the suburbs by making it easier for them to get to and from work. And, low-cost air travel has reduced costs, in time and money, of moving people over far greater distances. In addition, knowledge-based enterprises are far less dependent on movement of either raw materials or finished products. Most knowledge work can be delivered anywhere on the globe almost instantaneously at costs representing a very small fraction of its value.

Mind workers are more independent, even when they work for large organizations, and thus, require less frequent personal contact. For the first time in history, the link between a person's "workplace" and his or her home is being broken. People who continue to congregate around the old, large cities today do so more out of habit than out of necessity.

Naisbitt and Auberdene claim if today's cities didn't already exist, there wouldn't be any good reason to create them. Drucker claims that the city of the future might well become a center for information, communication, and entertainment rather than a center of work. "It might resemble the medieval cathedral where the peasants from the surrounding countryside congregated once or twice a year on the great feast days; in between it stood empty except for the learned clerics and its cathedral school."[12]

People are abandoning the cities for the suburbs for quality of life reasons: lower crime rates, quality housing as a lower cost, and recreational opportunities. Many people are now abandoning the suburbs for rural areas, for quality of life reasons as well – more living space, a cleaner environment, prettier landscapes, and perhaps most important, to regain a sense of community, a sense of belonging. Rising energy costs will raise costs of living in both cities and suburbs far more than living in self-sustaining rural communities. We currently label the urban to rural movement as "urban sprawl," but a better label might be the "resettling of rural America." Public policies addressing urban sprawl should be focused on how best to resettle rural areas, not on how to force people to continue living in overcrowded cities and overly expensive suburbs.

Dee Hock, founder of Visa International, concludes, "We are at that very point in time when a 400-year-old age is dying and another is struggling to be born – a shifting of culture, science, society, and institutions enormously greater than the world has ever experienced. Ahead, the possibilities of the regeneration of individuality, liberty, community, and ethics, such as the world has never known, and a harmony with nature, with one another, and with the divine intelligence such as the world has never dreamed."[13]

Countless books and articles have been written by hundreds of authors since the mid-90s trumpeting the "information age, the Internet era, the age of technology, or the new economy." However, most people had never even heard of the Internet – let alone used it – when the books cited here were written. The microcomputer had been talked about a good bit during the 1980s, but very few ordinary people owned one prior to the mid-90s. Biotechnology was discussed around agricultural colleges and medical schools, but GMOs or genetically modified organisms, were not the subject of ordinary conversations. When these futurists were writing in the late 1980s and early 1990s, the economy was still in the doldrums of lingering industrialization.

Obviously, many of the things envisioned by the futurists of the early '90s have not yet come to past, but their record for insights, thus far, has been pretty impressive. Meanwhile, the records of those who simply extrapolated past trends into the future, using mountains of data and sophisticated computer models, have turned out to be dead wrong. The world is being transformed into something fundamentally different. It takes human insight and intuition to assimilate the thousands of simultaneous relationships and to integrate the enormous detail and dynamic complexities involved in this world-changing transformation. However, such insights and intuitions are not the exclusive realm of the futurist. We all have the ability of see this new vision of the future; we need only learn to rely on our intelligent insight - our common sense.

Unfortunately, many people are distracted by current changes in technology and fail to understand that new technologies are but a reflection of our changing worldviews and operational paradigms. Industrial technologies didn't cause the industrial revolution but were a reflection of a revolution that took place first in the human mind. The idea of a fundamentally different way of doing things came first, and then innovative, creative, thinking people developed the technologies needed to implement these new ideas.

Machines were developed because someone first conceived how greater productivity could be derived from precise, repetitive processes, which were difficult for humans to perform. Factories were built because someone conceived of the potential that might result from organizing people and machines in highly standardized, linear, sequential relationships. Corporations were an old idea, but they were used in new ways to accumulate the capital needed to achieve the economies of large-scale, industrial production. The technologies that support a revolution may be new or old, but the revolution comes first to the hearts and minds of people.

Some have labeled the emerging post-industrial era as the "information age" - a time during which information will replace raw materials and the means of production as the source of new power and wealth. However, information creates new wealth only when it is used to do something productive - when it allows something of value to be created that would not have been created otherwise. The information revolution is not so much a matter of either creating new information or making existing information more readily accessible, it is much more about who has access to information, how information is to be used, and for what purpose. If new and better information is used by corporate decision-makers to maximize corporate profits, and they most certainly will keep such information from others if they can, then corporatism will continue. The new technology will simply support the old industrial system. However, if more people now have quicker access to more and better information, and use it to better their quality of life, then new information technologies will support the new revolution in human thinking.

For the new information technologies to be used for the common good, they must remain readily accessible to all - not just to those with power and wealth. One means of maintaining free public access to technologies, such as the Internet, might be to tax sales of private goods and services made over the Internet to generate funds to support and expand the internet system. Regardless of how it is accomplished, if information is to contribute to continuing human progress, it must reach the maximum number of human minds, everywhere. The new revolution is not embodied in new technology; change must come first in the hearts and minds of people.

It's interesting to note that the major innovations in new electronic information technologies were brought about by people outside of the old, industrial organizations. The microcomputer was not developed by IBM, but instead by a bunch of kids fooling around in their basements and garages. The Internet didn't blossom because of corporate communications but instead because a bunch of academics desired to share information about their research. The "dot.com" companies that fueled the "new economy" were nearly all started by corporate outsiders. These companies were not organized according to the old hierarchical corporate structures and they didn't usually manage by command and control. Certainly, these innovators sell their information products and technologies to corporate buyers and have become millionaires and billionaires in the process. And the large corporations are now scrambling to find ways to use these new technologies to their advantage, or at least to minimize its threat to the corporate status quo. But, the new information technologies were not developed to support industrialization, and thus, are not particularly industrial in nature. They were developed by people who had a different worldview and different vision for the future.

Inevitably, the initial "dot.com" investment bubble became over-inflated and burst. Skyrocketing stock values cannot be sustained indefinitely without actual profits. The information companies that survived the bubble burst were companies that actually facilitate increased productivity by providing something of unique value to people. And, most of these successful companies are fundamentally different from industrial organizations, because information is a fundamentally different kind of product.

These new information technologies hold tremendous potential for empowerment, if they can be kept out of the greedy grasp of corporate control. One of the basic motives for industrialization has been to maximize the value of scarce information. Industrial corporations, including government bureaucracies, have little to gain and much to lose if common people gain quick and easy access to vast amounts of practical and useful information and become willing to share it with others.

Over the past decade, I have had the privilege of working closely with one group of people with a different worldview and a different vision for the future. I call them the "New American Farmers." I have met with groups of farmers, from fifteen to fifteen hundred, all across the U.S. and in several provinces of Canada addressing the issues of agricultural sustainability. I never forego an opportunity to talk with a farmer. Through these conversations, I have been able to share in the emerging vision of the new American farm.[14]

While there are no "blueprints" for this new way of farming, some basic characteristics are emerging. First, these farmers see themselves as stewards of the earth. They are committed to caring for the land and protecting the natural environment. They work with nature rather than try to control or conquer it. They fit the farm to their land and climate rather than try to bend nature to fit the way they might prefer to farm. Their farming operations tend to be more diversified than conventional farms, because nature is diverse. Diversity may mean a variety of crop and animal enterprises, crop rotations and cover crops, or managed livestock grazing systems, depending on the type of farm. By managing diversity, these new farmers are able to reduce their dependence on pesticides, fertilizers, and other commercial inputs that squeeze farm profits and threaten the environment. Their farms are more economically viable, as well as more ecologically sound, because they farm in harmony with nature.

Second, these new farmers build relationships. They tend to have more direct contact with their customers than do conventional farmers. Most either market their products direct to their customers or market through restaurants or specialty retailers, sometimes with agents helping them to develop relationships with their customers. They realize that as consumers each of us value things differently because we have different needs and different tastes and preferences. They produce the things that their customers value most, rather than try to convince their customers to buy whatever they produce. They are not trying to take advantage of their customers to make quick profits; they are trying to create mutually beneficial relationships. They market to people who care where their food comes from and how it is produced – locally grown, organic, natural, humanely raised, hormone and antibiotic free – and, they receive premium prices by producing things their customers value most. Their farms are more profitable as well as more ecologically sound and socially responsible.

These new farmers challenge the stereotype of the farmer as a fiercely independent competitor. They freely share information and encouragement. They form partnerships and cooperatives to buy equipment, to process and market their products, to do together the things that they can't do as well alone. They are not trying to drive each other out of business; they are trying to help each other succeed. They refuse to exploit each other for short run gain; they are trying to build mutually beneficial relationships. They buy locally and market locally. They bring people together in positive, productive relationships that contribute to their economic, ecological, and social well-being.

Finally, to these new farmers, farming is as much a way of life as a way to make a living. They are "quality of life" farmers. To them, the farm is a good place to live – a healthy environment, a good place to raise a family, and a good way to be a part of a caring community. Their "quality of life" objectives are at least as important as the economic objectives in carrying out their farming operations. Their farming operations reflect the things they like to do, the things they believe in, and the things they have a passion for, as much as the things that might yield profits. However, for many, their products are better and their costs are less because by following their passion they end up doing what they do best. Most new farmers are able to earn a decent income, but more important, they have a higher quality of life because they are living a life they love.

My brother Don is one example of the new American farmer. He took over the small dairy farm where I grew up, after my father died in the mid '60s. Over the years, he tried to increase production per cow and build the size of his herd – to become a "good" dairy farmer. Sometime in the early '90s, however, Don decided he was going to have to try something radically different. He needed to find ways to cut costs rather than continually try to increase production. He began a transformation to a grass-based system by relying far more on pasture and hay and far less on corn silage and grain supplements. He developed his own approach to "management intensive grazing" – rotating his cows among different pastures to maximize forage production and quality. He developed his system gradually, but by the end of the '90s, Don's cows were moving through thirty-some paddocks of pasture that included the whole farm.

Over the decade, he had cut back from milking close to 100 cows down to less than half that number, and milk production per cow had dropped as well. Although his total milk production had dropped by more than half, his costs of production had dropped even more dramatically. Today, he is milking fewer cows, has quit feeding corn silage, and as a result, has much lower equipment costs and is doing much less work on the farm than before. And equally important, he told me in the late 1990s that he had made more money each year, year after year, ever since he had started trying to reduce costs rather than to increase production.

Don's farm has its ups and down, as do all farms. But, of the twenty-plus dairy farms that once lined the surrounding roads, my brother's is one of only two or three left. He has been able to sustain a desirable quality life on the farm while his neighbors were being forced out of the dairy business, because he was willing to search for and try a new and better way of farming.

There are literally thousands of these new farmers all across the continent, creating their own versions of new and better ways to farm. They are on the frontier of a new and different kind of agriculture, an agriculture that is capable of meeting the needs of the present while leaving equal or better opportunities for those of the future – a sustainable agriculture. These farmers face struggles and hardships and there are failures along the way. Life is rarely easy on any new frontier. But, a growing number are finding ways to succeed.

Sustainable farming is thinking farming. It requires an ability to translate observation into information, information into knowledge, knowledge into understanding, and understanding into wisdom. Certainly, sustainable farming involves hard work, but farming sustainably is not the first stage of development beyond hunting and gathering, as some historians might label it. It is the next stage of development, beyond industrialization. Sustainable agriculture is very much in harmony with a post-industrial paradigm of development, the next step forward in the ongoing process of human betterment. Sustainable farmers are thinking workers, or working thinkers.

I know from contacts in both the U.S. and Canada that similar "movements" are taking place in natural resource management – forestry, fisheries, wildlife – in healthcare, education, and international development. All represent a rejection of the industrial model because it is socially exploitative and ecologically destructive. All these movements advocate a more holistic, diverse, individualized approach to their respective professions. In 2002, the United Nations held its Third International Summit on Sustainable Development, during which the United States was consistently labeled a "laggard" for its reluctance to respond to the issue of global sustainability. Thus, the concept of sustainability is embraced even more openly in other parts of the world than in the U.S.

The sustainability movement is completely consistent with the visions of Toffler, Senge, Drucker, Naisbitt, Reich, and others who see knowledge and information as the keys to future human progress. The new farmers, for example, manage their unique, complex systems of production to meet the unique needs of specific customers while protecting the natural environment and nurturing positive personal relationships. These new farmers are joining the ranks of the mind-workers of the future. However, information and knowledge alone cannot sustain human progress on earth. We are more than our bodies and we are more than our minds; we also have souls. The new era of human progress will not rely on logic and reason alone, but must also rely upon people using their intelligent insight – their common sense.

This chapter has provided but a few glimpses of new visions for the future. I have used them to present my vision of hope for the future. One vision of hope may be enough for one person, but it is not enough to start a revolution or to bring about a fundamental transformation of the world. One vision of hope can change a life, but a shared vision of hope can change the world. Shared hope is the product of communication, one person tells another about their vision, and that person shares their vision in return. No two visions will be the same, but many visions will have some elements in common; those held in common reflect our common sense of hope.

As more and more people share their visions with others, the elements that are widely held in common, the shared hope, will become evident. And, as we see that some of our hopes are held in common among other thoughtful, intelligent people, and we know this hope is coming from different sources, we may infer that other reasonable people have seen the same things as we. We know that we haven't been dreaming. We know that our shared hope for the future can become a reality, if we have the courage to make it so.

As we begin to build bonds of shared hope with other like-minded people, the strength of our shared vision grows – not just by some multiple or exponential, but it literally explodes as a living, growing epidemic of hopefulness. In our shared vision of a better future, there is a realistic hope for victory. All we need do is rely on the sense we share in common, on our common sense.

Endnotes

[1] Robert Pool, "Science Literacy: The Enemy is us," *Science*, (1991), 266-267.

[2] Elizabeth Culotta, "Science's 20 greatest hits take their lumps," *Science*, (1991) 1308-1309.

[3] Alvin Toffler, *Power Shift* (New York: Bantam Books, Inc. 1990).

[4] Robert M Pirsig, *Zen and the Art of Motorcycle Maintenance* (New York: Bantam Books, Inc., 1974).

[5] Pirsig, *Zen and Art*, 241.

[6] Peter M. Senge, *The Fifth Discipline* (New York Currency Doubleday, 1990).

[7] Peter Drucker, *Post-Capitalist Society* (New York: HarperBusiness, 1994).

[8] Drucker, *Post-Capitalist Society*, 1.

[9] Peter Drucker, *The New Realities* (New York: Harper and Row Publishers, 1989), 259.

[10] Robert B. Reich, *The Work of Nations* (New York: Vintage Books, Random House Publishing, 1992).

[11] John Naisbitt and Patricia Aburdene, *Megatrends 2000* (New York: Avon Books, 1990).

[12] Drucker, *New Realities*, 259.

[13] M. Michell Waldrop, "The Trillion-Dollar Vision of Dee Hock," *FastCompany.com*, October, 1996 <http://www.fastcompany.com/magazine/05/deehock.html> (accessed September 2006).

[14] John Ikerd, *Crisis and Opportunity in American Agriculture: Essays on Sustainable Agriculture*, (Lincoln, NE: University of Nebraska Press, 2008).

12

SOWING SEEDS OF REVOLUTION

In a telephone conversation in 1995, I was talking with a representative of a grass roots organization of farmers in north Missouri, trying to recruit the organization as a partner in one of our sustainable agriculture programs. Near the end of our conversation, I asked the representative if I could do anything else for his organization. He replied something to the effect, "You can burn that paper your department just put out supporting large scale confinement animal feeding operations (CAFOs) as a rural economic development strategy." I told him I was philosophically opposed to burning any kind of printed information, regardless of whether I agreed with it, but I would take a look at the paper and get back to him about it.

He sent me a copy of the paper and I read it. It was written by one on my colleagues. He had conducted an economic analysis showing the numbers of jobs that could be created in Missouri, if rural communities would support these large-scale "animal factories." CAFOs are the epitome of industrial agriculture. Thousands, often hundreds of thousands, of animals are crowded together in large enclosed buildings, with the animals confined in spaces that hardly allow them to turn around. Production in these facilities is literally a manufacturing process, where tons of feed come in one end, mountains of manure are flushed out the other. The animals are "manufactured" on a biological assembly line, as they consume the feed and excrete the waste. These operations are specialized, standardized, and consolidated under the control of giant agribusiness corporations, typically through comprehensive contracts under which contract growers become little more than corporate serfs.

Some of these CAFO operations generate more biological waste than do large cities, and the manure typically is stored in open pits and spread openly on surrounding fields. The manure often creates offensive odors for miles around, pollutes streams and groundwater, and creates significant health risks for residents of the local community, eventually prompting a call by the American Public Health Association for a moratorium on further construction.[1] It's no surprise that a new CAFO locating in a rural area creates instant dissension and disruption within the community, between the few who will benefit economically and the many, particularly those downwind and downstream, who will suffer the environmental consequences.

I reviewed the paper and immediately started to write a rebuttal. In summary, I concluded that for every job created in a contract CAFO operation three independent family hog farmers likely would be displaced. New CAFOs do nothing to increase consumption of meat, so every pound of meat they produce reduces the market for meat from another producer somewhere. I relied on University of Missouri's farm records program to determine the number of independent hog farmers currently employed to produce the number of hogs that would be produced by the CAFO operations suggested in the paper. I concluded that a given number of hogs would employ three times as many independent hog farmers as CAFO employees, with no significant differences in costs of production or earning of the two groups of people employed. If communities wanted to produce more hogs, then CAFOs made sense, but if communities were interested in jobs for people, they needed to support their independent family hog farmers.

Needless to say, the paper created quite a furor – not just in my department, but also throughout the state and across the nation. It eventually was published in two different books.[2] The paper offended the corporate agricultural establishment, and the University of Missouri would have fired me, if they could have found any way to do it without ending up in a lawsuit. My academic freedoms were threatened so many times I lost count, and I still have the documents to prove it. Since writing that paper, I have met with rural community members fighting CAFOs in more than 15 states and three provinces in Canada. Among these community activists, I see the beginning of a new social and political revolution.

The community activists fighting CAFOs are among the new economic, social, and ecological revolutionaries. They are seeing firsthand how the pursuit of short run economic self-interests destroys the ecological and social resources upon which the long run economic viability of their communities must depend. They understand that the pursuit of greater wealth by a few is destroying not only the future of their communities but also the overall quality of life within their communities today. They understand that the future of rural America depends upon finding systems of economic development that are also ecologically sound and socially responsible and thus are sustainable.

Perhaps most important, these concerned and thoughtful rural people are beginning to stand up for their basic rights as members of a democratic society. These activists are being opposed by some of the most powerful of economic and political interests in the world. They have little influence at federal and state levels, because agricultural and environmental policies at those levels are controlled by the multinational corporations. So they are learning to claim their basic rights as individuals at the local level, within their own communities, where they cannot be so easily ignored. They are educating and informing themselves by sharing information with other local groups all across the continent, who have fought or are fighting the same battles. The internet has become a power tool, not only to share information but also to locate the latest research results related to CAFOs, and to keep abreast of legislative and regulatory developments. They are organizing, holding rallies, lobbying their county commissioners and state legislatures, and otherwise becoming good citizens. And they are standing up for their democratic rights of self-defense and self-determination – the basic rights of all people to protect their health and natural environment against economic exploitation.

It's time for this same kind of social and political revolution to spread all across America. Our democratic society is dying a slow and painful death. If we don't act boldly, decisively, and quickly to restore it, we will lose it – forever. Democracy, by its very nature, requires active participation by individual citizens. Some Americans still participate in the political process but, for the most part, they participate impersonally, through special interest groups, through labor unions, through business organizations – through various kinds of "corporations." Our democracy has degenerated into a corporate aristocracy – a government in which corporations rule, not simply because they control the nation's wealth, but also because they are considered to be superior in their intelligence, organization, and culture. We have become a nation that allows our government to be "run like a business." The real people, the common people, have in effect renounced their citizenship, leaving the powers of government in the hands of the new aristocrats – the leaders of large corporate organizations.

Alexis de Toqueville, in the early 1800s, warned of the tendency of democracies to concentrate political power in their central governments. But, he also warned of the potential for new aristocracies to emerge from within democracies in the form of "manufacturing organizations" – that is, corporations.[3] However, he apparently did not anticipate that these new corporate aristocracies would eventually seek control of the strong central government. He wrote, "The aristocracy created by business rarely settles in the midst of the manufacturing population which it directs; the objective is not to govern that population, but to use it."[4] He apparently did not perceive that industrial corporations eventually would not only exploit people as workers and consumers but would exploit them as taxpayers as well.

Thus far, most Americans seem to accept the political and economic degradation of our society, because they quite simply do not understand what is happening. They sense that something is wrong, but they are continually reminded that they have more stuff than any society in human history. They are chastised for their ungratefulness and are threatened with deprivation if they question the means by which our material affluence has been achieved and maintained. They are made to feel unpatriotic if they question the legitimacy of strategic wars that are obviously being fought for control over global resources. They are told that science – logic and reason – supports our current economic and political systems, and to oppose them is illogical and unreasonable. In response, most have closed their eyes, even to the evils they know to exist.

Those who question the common sense of our current economic and political systems are labeled as emotional and irrational, or as misguided activists and radicals. People are shamed into participating in an exploitative economy and oppressive society because they are afraid to challenge the intellectual, political, and economic authority of "the establishment." Perhaps equally important, many people are afraid they cannot survive economically without the corporate aristocracy. They depend on the corporations for their employment, their fringe benefits, their retirement, the local tax base, and their overall economic well-being. They have become slaves of their economic dependency.

However, more and more people are beginning to wake up to the fact that our survival is threatened by the very existence of giant corporations, not by the prospect of their demise. A life of economic slavery can never be a life of economic security. Economic security must come from our ability to care for ourselves and to contribute something of value to others, not from corporate promises of some greater future prosperity.

We need to remind ourselves that as we move into the new post-industrial era of economic and social development, knowledge, not capital, is the most important of our productive resources. Human imagination and creativity are products of individual thoughts and of interactions among individuals within communities of thinking people. We must have an economic system that empowers people, rather than enslaves people. In the new economy, there will be little economic justification for publicly owned corporations because there will be little need to accumulate large amounts of capital or large numbers of people to achieve economies of large-scale production. The most efficient and productive organizations in the future will be small, dynamic, living – not large, rigid, and industrial. The industrial era is over. It's time to dismantle the archaic corporate structures of industrialization and to replace them with an infrastructure that will support the new era of sustainable human progress.

People today are perfectly capable of making their own decisions; they don't need some large corporate entity to make their decisions for them. Today's society is admittedly complex, but new information technologies have eliminated the need for people to make their political decisions through special interest groups, that is, through corporate intermediaries. All of the information people need to make informed decisions on any political issue is already available – through electronic media, the print media, mass mailings, word-of-mouth, or the Internet. People need only spend the time and effort necessary to access it.

Certainly, people still need representatives to carry out the day-to-day functions of government. But even in a representative democracy, people must accept their responsibility of participating in the selection of their representatives and of informing those who represent them of their positions on public issues. There may be a continuing role for organizations in assembling and distributing relevant information on specific issues, but people no longer need others to make their decisions.

Through interactive communications, we even have the means as individuals to carry on a national dialogue, to move toward a national consensus concerning how to reshape our economy and restructure our government to meet the changing needs of the people. We have the means of restoring our democracy, restructuring our economy, and recreating our government. All we need is the courage to act.

The logic and reason supporting our current economic and political systems are no longer the only logical and reasonable systems of logic and reason. A new philosophical foundation had been laid upon which we can build new ways of thinking and of reasoning. Quantum physics and chaos theories, for example, are logical, reasonable alternatives to the old theories of mechanical physics and statistics. These new concepts, in a sense, redefine the concepts of logic and reason, recognizing explicitly the interconnectedness of all things and the dynamic complexity of the living world. These new ways of thinking and knowing validate the value of human insights, intuition, and judgment, both in science and in day-to-day life. These new insights into the nature of reality validate our continuing confidence in using our common sense in addressing the most important issues of the future.

The future progress of humanity will not be measured by our continued accumulation of still more cheap stuff, but instead will be measured by our willingness and ability to use wisely those things we have.
Nor will our future progress be measured by our greater independence,
but instead, by our willingness and ability to use our independence more
wisely. Future human progress will be measured in terms of the quality
of our relationships – our relationships with each other and with nature.
We will learn to rely on each other as a matter of choice, not out of
necessity, but because personal relationships make our lives better. We
will learn to "walk lightly on the earth" by choice, not out of necessity,
but because doing so makes our lives better. We will find greater sense
of purpose and meaning in our lives as we develop and nurture positive
relationships with each other and with the earth. We will find a higher
quality of life as we create a more sustainable human society and this
will be the measure of future human progress.

As we return to our common sense, we will learn to listen again to
the voice of intelligent insight within. Our common sense tells us that it
is fundamentally wrong to exploit other people and to exploit the natural environment in pursuit of our narrow, individual self-interests. Our
common sense tells us that our quality of life is enhanced by positive
relationships with other people, regardless of whether we receive anything of economic value in return. Our common sense tells us that our
quality of life is enhanced by our stewardship of natural resources for
the benefit of those of future generations, although we know that we
receive nothing of economic value for our efforts. Our common sense
tells us that taking care of our individual self-interests is necessary but is
not sufficient to ensure a desirable quality of life. Our common sense
tells us that we must pursue a broader, higher, more enlightened concept
of self-interest. We don't have to prove these things to others or to ourselves – we know them, they are plain common sense.

Our common sense also tells us that it's time for fundamental economic change. An economy is an organization created by people for the
benefit of people. It doesn't make sense for people to remain slaves of
an economic system of their own creation. Our common sense tells us
that it's time for fundamental political change. The government is an
organization created by people for the benefit of people. It doesn't make
sense for people to remain slaves of a political system of their own creation. We created our economy and our government and we can recreate our economy and our government, any time we choose. We have the
power to change the things that oppress us, if only we can find the
courage to act.

We must find the courage to restore government so that it serves the
common good. The private economy will not provide those things to

which people have "equal" rights in their pursuit of happiness, with "equity and justice for all." The private economy rewards in relation to a person's ability to contribute, not in relation to their needs or their rights. Government is the only means by which we can ensure equity and justice. Government also provides the means by which we can do many things more effectively and efficiently by doing them collectively rather than individually. Such collective decisions are matters of choice. But in matters of equity and justice, we have no choice but to act collectively, through government. If we are to live in a civil society, we must reclaim our government.

If we are to create a sustainable society, we must change our economic ways of thinking. We must develop an economic system that considers the social and ecological as well as the individual implications of our decisions. In the individual economy, decisions are appropriately based on economic values, dollars and cents, because the costs and benefits of such decisions accrue almost exclusively to the individual. But, in the social economy, each person must be given an equal voice in the decision making process, because each person has an equal right and responsibility to participate for the benefit of the common good. And, in long run ecological and social economies, decisions are ethical or moral in nature; they affect the well-being of those of future generations who can neither buy or sell in today's market place nor vote in today's elections. Thus, decisions in the moral, ecological economy must be made by the process of consensus concerning our ethical responsibilities to care for the earth – to ensure the rights of future generations. We must remake our economy to enhance the individual, social, and ethical qualities of our lives.

People may say, "If it ain't broke, don't fix it." I agree that the current system may not actually be broken, but it is rapidly breaking down, and if we wait until it breaks, we may not be able to fix it. We may not be able to restore health to our economy, if we wait until it collapses and millions are jobless and homeless. We may not be able to restore health to our ecosystem, if we wait until we have depleted our biological and energy resources, or we have created a global climate in which we humans can no longer survive. We may not be able to restore health to our society, if we wait until global society is so splintered into irreparable ethnic, socioeconomic, and religious factions that global nuclear conflict becomes inevitable. Even if it ain't broken yet, now is the time to fix it. If we must break it to fix it, then let's break it now, while it can still be repaired.

The world will continue to change, regardless of what we do or don't do. But by our refusal to act now, we may sentence the whole of humanity to eventual economic and social chaos, returning civilization to another "dark ages." By our refusal to act now, we may plunge the world into violent revolution against the global corporate oppressors – into a never-ending war on terrorism and global military conflict. We may prove that Marx was right after all, that the wealthy capitalists are destined to continue exploiting the working class until the workers revolt and overthrow their governments. Or by our refusal to act, we may be sentencing humanity to a long, slow death, slowly draining the earth of its social and ecological resources and surely destroying any hope for the future of human life on earth. Now is the time to act, while there is still time to act. If we tarry too much longer, it will be too late.

Some actions we can take alone, but others we must take together, through government. Even if one can only view government as a "necessary evil," there has never been a time when government, even if evil, was more necessary. As Thomas Paine put it in his essay, *Common Sense*, "Society is produced by our wants, and government by our wickedness; the former promotes our happiness positively by uniting our affections, the latter negatively by restraining our vices."[5] He concludes, that government is "rendered necessary by the inability of moral virtue to govern the world."

We must have government to ensure equity and justice. People inherently lack the "moral virtue" to ensure these inalienable rights without the authority of government to interpret and enforce their "moral consensus." We have a national consensus that murder and robbery are wrong, for example, yet we need the government to protect us from murderers and robbers. In the current age of moral and ethical decay, the civility of our society is critically dependent on an effective, fair, and just government. There has never been a time when government was more necessary, even if some may see it as inherently evil.

The tendency of democracy toward a strong central government is a real risk. Government tends to work best when it is decentralized, when decisions are made closest to the people affected by those decisions. Local involvement and local control remains the cornerstone of democracy. But a strong central government is absolutely necessary to "restrain the vices" of today's giant corporations. Corporations have no "moral virtue" to advise them, and thus, are completely incapable of restraining themselves.

State and local governments will have the rights to protect their citizens from corporate threats to their health and natural environment, only if the federal government uses its constitutional authority to ensure those rights. In fact, we may need to develop some effective form of international government to restrain multinational corporations, in spite of the inherent risks of world government. We simply have no means other than government of confronting the corporate aristocracy. Once we have dealt with the real tyranny of corporatism, we can then return our attention to dealing with the potential tyranny of a strong central government.

The problem today is not that our government is too strong, but instead that it is too weak. It has lost its power to "say no" to corporate demands. The government has become an impotent partner in its economic intercourse with the corporate world. In fact, we no longer even have checks and balances among the legislative, executive, and judicial branches of government. All three branches are under strong corporate influence, and on many issues, are under corporate control.

Congress may disagree with the President on issues such as civil rights, education, or crime prevention. But the two branches of government never disagree on issues concerning the priority given to the economy over the people. They may argue about whether a new law to protect the environment or to ensure social welfare will result in a significant drag on the economy, but there is never any disagreement over whether the economy should take precedent over the social welfare of the people – of either current or future generations. The President and Congress agree on the conventional fallacies that a strong American economy is the most important priority of government, and that strong American corporations are essential to a strong American economy. Those who would challenge these fallacies will not be elected President, and only rarely will be elected to Congress; of this, the corporations make certain.

The Supreme Court is supposed to provide the ultimate balance of power in Washington. They are charged with interpreting the Constitution – the consensus of the people – concerning issues of ethics, morality, freedom, and justice. Even at the highest court in the land, however, the corporations are awarded greater rights with fewer responsibilities than are real people. The courts have given corporations human rights but without human responsibilities. When new Justices are nominated to fill vacancies on the Supreme Court, invariably questions arise concerning their views on abortion – are they "pro-life" or "pro-choice" – and their positions on civil rights, gun control, and states' rights. But questions concerning whether they are "pro-economy, pro-business, or pro-corporate" never seem to arise. Those who give a higher priority to social equity and justice than to ensuring a strong economy, quite simply are never nominated; of this, the corporations make certain.

Some may argue that the corporations have been granted certain rights by their charters, that the Supreme Court has upheld these rights, and thus it would be unfair to investors for the people to infringe upon on these rights, and certainly unfair to take them away. However, as Thomas Paine argued years ago, no generation of people can give away the rights of any future generation, whether the right of self-determination, ceded to a monarchy, or the right to consolidate market power or influence a political election, ceded to a corporation. The people of this generation have a fundamental right to throw off the yolk of corporate oppression - regardless of what rights may have been granted by previous generations. Those who have invested in corporations have no more right to compensation than did those who were counting on continuation of the British Monarchy in the days of the American Revolution.

The people of this generation have a responsibility to restore democracy - to return control of the economy and the government to the people. No past generation had the right to trade away the economic and political freedom of this generation, or of any generation of the future. It's time to take back those freedoms, even if it takes a revolution.

The current government seems incapable of acting, so we must act for ourselves. We have no checks and balances, because all branches support the same economic agenda - an agenda for exploitation of people and of nature. The people are the only remaining checks and balances, for either the government or the economy. So, it's up to us to restore integrity to both, because neither currently works for the common good. Our common sense tells us that it's time for the real people to take control.

The "free market" has become the new monarch, but it's time to dethrone the king. We have blindly accepted that "whatever is profitable shall be done," regardless of the ecological or social consequences. It's as if God had decreed that the "markets work," and none should question this wisdom. As Thomas Paine pointed out in *Common Sense*, people fall into error whenever they allow themselves to interpose a king so as to avoid having to make moral judgments for themselves. Paine quotes from the scriptures where the Jewish people asked Samuel to "make us a king to judge us, like all other nations." Samuel thought they wanted a king because they were displeased with his leadership. So, "Samuel prayed unto the Lord, and the Lord said unto Samuel, 'Hearken unto the voice of the people, for they have not rejected thee, but they have rejected me, that I should not reign over them.'"

We have made the "free market" our king, so that our moral and ethical values - our common sense of right and wrong - should not reign over us. "The markets" will transform our greed into good, we reasoned, so we no longer had to be concerned with judging what was right or wrong for ourselves. We were free to pursue our self-interest, without regard for other people or concern for God's earth. We had found a benevolent king – so we believed – to judge for us. But, it's now clear to all "who have eyes to see" that our monarch, the "free market," is destroying the earth and its people. It's time to dethrone the king and to restore self-rule, through our common sense of ethics and morality.

It's time to accept the personal responsibility for making moral and ethical choices, rather than simply to accept the impersonal judgement of the marketplace. It's time for a revolution of ethics and morality – not just in our personal lives, but in matters of business and politics as well. The free market is not equitable and the economy is not just. The people must supply the necessary ethical and moral constraints to individual and corporate greed. Individual greed may be a personal matter, but corporate greed is a public matter that can be addressed only through government. Therefore, we must restore morality in government as well as in our personal lives.

We must reject the conventional wisdom that the "American Dream" depends on a strong corporate economy. We must reject the daily barrage of public and private propaganda that the jobs and incomes that support American families and the tax base that supports the U.S. government are all dependent on an ever-growing corporate economy. The truth of the matter is that most new jobs, at least for the past few decades, have been created by small entrepreneurial organizations. The corporations have been exporting good paying American jobs to other countries, where wages and benefits are far lower. Most of the new corporate jobs are low paying, part-time jobs, with limited benefits. We are witnessing the destruction of middle-class America. The future of employment in America cannot be trusted to corporate interests. Corporations do not account for a large portion of taxes collected and the corporate share of the tax burden seems to fall each year. The vast majority of all taxes collected come from individuals – a disproportionate share coming from the dwindling American middle-class. Our economy and our society are supported by real people – not by corporations.

We also must reject the conventional wisdom that our government should be "run like a business." This too is propaganda peddled by the corporate community, to ensure that government continues to serve their interests rather than the interests of the people. Everyone has an equal right to public goods and services regardless of whether they contribute anything to tax revenues. The role of government is not to minimize costs, but to ensure the provision of public goods and services, as efficiently and effectively as possible without compromising its public purpose. The government must do those things that need to be done, but that the private economy won't do. The benefits of doing such things, and the costs of not doing them, cannot be measured by the yardstick of economic efficiency alone.

We must reject the conventional wisdom that the people are powerless to change either their government or their economy. We have the power to elect a different group of people to send to our state capitols and to Washington DC, with a mandate to manage the economy for the common good. The corporations control the election process today only because we let them. We continue to relinquish our power to elect people who will truly represent us only because we fail to act. We also have the power to amend the Constitution to protect the rights of "all people" against "economic oppression" – to ensure equality and justice for all. We can amend the Constitution to grant those of future generations these rights as well – to sustain a healthy, productive natural environment for all people of all times. Our Constitution ultimately must confirm that we have no more right to own something, just because we have enough money to buy it, than we have to steal something, just because we are physically strong enough to take it. People have an inherent right to decide what they will and will not sell, just as they have a right to decide what they will and will not give away. Economic oppression is no more moral or ethical than is physical oppression.

The fact that some past generation willingly sacrificed the basic human rights of some people for the economic benefits of other people does not absolve us of our responsibility to restore economic equity and justice. The fact that past generations may have exploited the earth, knowingly or unknowingly, does not absolve us of the responsibilities of our generation to restore the integrity of the earth for the benefit of those of the future. The Constitution belongs to the people. We are not just subjects of the Constitution, but are the creators and caretakers of the Constitution. We have both the right and the responsibility to change the Constitution to meet the changing needs of the people.

However, the revolution must begin within each of us. The economic and social revolution will succeed only if we first succeed in our individual, personal revolutions. For guidance, we need only look to Thomas Paine. First, we must clearly identify the villain or the enemy from which we are renouncing our allegiance or subjection, regardless of whether it is the villain within us or the world around us. Second, we must have a positive vision of hope for the future – a vision of life after we have broken the bonds of allegiance or subjection. And finally, we must be willing to commit our lives and our fortunes to a course of thought and action, never doubting our ultimate victory. This is the anatomy of a revolution – the villain, the vision, and the victory. This is the key to victory – individual, societal, and global.

I see this same pattern of revolution emerging among the people of rural America today. They are beginning to understand their villain - the corporate exploitation of people and nature. The most popular economic development opportunities offered remote rural communities today include prisons, landfills for urban waste, toxic waste incinerators, and giant confinement animal operations. The highest economic use the corporate economy places on rural America is as a place to dump the human, material, and biological wastes of their extractive, exploitative systems of production. It's not a rural versus urban issue; rural people simply have less political power to protect themselves from exploitation.

A few rural communities also are beginning to see a new vision for their future, through sustainable rural economic development. They are beginning to realize that their most valuable economic resources over the long run are clean air and water, open spaces and scenic landscapes, and caring communities within which people can find a sense of belonging and a desirable quality of life. As this vision becomes shared in common among more and more people, the revolution to reclaim rural America will truly begin.

Once the revolution begins, there will be no doubt of the ultimate victory. After a few communities demonstrate to urbanites and suburbanites that rural communities could be good places for them to live, they will find ways to make an economic living there. Innovative, creative, imaginative people can make a living anywhere they choose to live in the economy of today and even more so in the future. In addition, a few people can have far more influence on a small rural community than they can ever expect to have in a large city. A few rural revolutionaries can fundamentally change the future of rural America. Once Americans wake up and return to the pursuit of happiness, rather than wealth, the renaissance of rural America will be unstoppable.

As we in the larger society wage our personal revolutions, we will find the courage we need to wage the political revolution needed to reclaim our democracy and our economy. Eventually, we will develop a new national consensus and will rewrite the Constitution to protect the people from economic exploitation and to ensure the rights of those of future generations. Changes in the hearts and minds of the people eventually will be reflected in changes in the "laws of the land." Only then will this revolution be won. This is the nature of the new revolution. The time for revolution is now, before it is too late. Let it begin!

Endnotes

[1] American Public Health Association, *Association News*, 2003 Policy Statement, <http://www.apha.org/legislative/policy/policies.htm> (accessed September 2006).

[2] John Ikerd, "Sustainable Agriculture, Rural Economic Development, and Large-Scale Swine Production," in *Pigs, Profits and Rural Communities*, eds. Kendall M. Thu and E. Paul Durrenburger (Albany, NY: State University of New York Press, 1998).

[3] Alexis de Tocqueville, *Democracy in America*, (New York: Bantam Books, 2000, original copyright, 1835), 690.

[4] Tocqueville, *Democracy in America*, 693.

[5] Thomas Paine, *Common Sense* (Mineola, NY: Dover Publications, 1776, republished 1997).

13

RESTORING BODY, MIND AND SOUL:
THE PERSONAL VICTORY

We human beings are physical, mental, and spiritual beings. Thus, the personal victory will represent a triumph of the whole person – the spiritual, mental, and physical being. Perhaps the greatest human tragedy of the modern scientific era has been the triumph of the mental over the spiritual. In this age of reason, people have been purposefully persuaded to doubt, if not outright deny, their spiritual nature. Those who have resisted have been coerced into keeping the spiritual aspect of their lives to themselves. Spirituality has been systematically excluded from our schools and public affairs under the guise of separation of church and state. Spirituality has been discouraged in the workplace, under the guise of avoiding unnecessary distraction. Spirituality has been excluded from science, because it isn't considered to be logical and rational. If you can't prove that something exists, but you still know it exists, then you should just keep it to yourself. To achieve personal victory, we must reclaim the spiritual dimension of our lives.

Within a matter of weeks, in mid-1997, I received two requests for keynote presentations at agricultural conferences, both wanting me to address the spiritual dimension of farming. I was aware that spirituality had been a theme within the sustainable agriculture movement, in fact, a national conference on "The Soul of Agriculture" had been held in New England the past spring. But most of what I had read or heard about spirituality and agriculture thus far had been pretty academic in nature. I don't know where the people who invited me got the idea of asking me to speak on spirituality. Perhaps I had made some comments on the Internet during exchanges related to moral and ethical issues that had caught their attention. I honestly don't recall. But I accepted the challenge of preparing a presentation on the subject. The title I liked best was, "Reclaiming the Sacred in Food and Farming."

These two presentations, one in Wisconsin and the other in North Carolina, changed my professional career – and my life. Never before had my speeches sparked such enthusiastic responses. I received standing ovations from both groups. I was almost moved to tears by the positive energy coming back from those audiences. Members of the audience in both places came up afterwards and told me, it was as if my words had been coming through me but from somewhere beyond. I shared the feeling. The words truly had come from my heart, but they also had come to my heart from somewhere beyond. Before each presentation I had prayed, "God, please let me speak Your words, not mine – the words You want these people to hear. You have brought me to this place for a purpose, if I am to fulfill that purpose, you will need to give me the words you want me to speak." This certainly was not the first time I had prayed such a prayer, but it probably was the first time I had found the courage to do my part – to say what I felt God wanted me to say.

Now, whenever I am asked to make a new presentation, I generally prepare a paper on the topic of my talk, but I never simply read a paper to the audience. When I speak, I rely on small note cards on which I have a brief outline of the points I want to make. I review my paper and my cards before every presentation, but I then depend on God to give me the actual words that come out of my mouth. I joke with people that I never know what I am going to say until I have already said it, and to some extent, this is true. I am learning to listen to the voice of my soul.

The term "soul" is sometimes used to identify the essence of a person, including the mental, physical, and emotional or spiritual. However, when I refer to the soul, I am referring to the spiritual aspect of a person, thus distinguishing among the body, the mind, and the soul. When I speak publicly about spiritual matters, as I frequently do, many audience members seem surprised, if not shocked. Scientists are just not supposed to talk about such things, even if they are social scientists. We can talk about spirituality in the abstract, but we are not supposed to talk about our own spirituality or the relevance of spirituality to real issues in the lives of those in our audiences. However, I have never had a negative reaction from an audience member after I have addressed spirituality as an essential aspect of sustainability. I am careful to distinguish between spirituality and religion; in fact, I have had Christians ask me if I am a Christian, after I have given speeches with strong spiritual themes. Whereas people are reluctant to discuss religion, I have come to believe that a vast majority of people are hungry for opportunities to discuss spirituality.

All human beings have souls, regardless of our doubts or denials. Perhaps in some sense, we share a common soul; nonetheless, we each have a soul. We can never expect to understand fully the concept of soul, at least not in a rational and logical sense. Logic and reason are mental, whereas, the soul is spiritual. Our minds are reluctant to acknowledge the existence of anything that we can't understand through logic and reason. So our minds, being logical and rational, also encourage us to deny, or at least to doubt, the very existence of our soul.

Even in organized religion, the concept of sin has been reduced to the struggle between mind and body. We are considered righteous if we are able to resist the temptations of the flesh – if we don't kill anyone, don't steal anything, don't lie to gain advantage, don't commit adultery, and don't lust for things that belong to others. The Old Testament *commandments* not to worship false gods or to deny the existence of the one true God have become optional *suggestions* in many religious circles today. After all, is it reasonable to expect even religious people to believe in things that their minds can't logically understand?

Barbarians allow the animalistic urges of their bodies to control their actions, but civilized people put their minds in control of their bodies. Why should we expect civilized people in the age of reason to allow their minds to be influenced by some intangible, mystical, spiritual idea called the soul? But intelligent, civilized human beings do allow their minds to be influenced by their soul. They know they have souls and they know that only the soul is capable of controlling many of the most important aspects of their lives. To deny what they know to be true is to be truly unreasonable.

The human mind quite simply is not capable of discovering the purpose or meaning of the life of which it is a part, at least not through reliance on the scientific method or any other logical method of reasoning. The mind can't possibly discern why we humans benefit from the simple act of treating other people as we would have them treat us. The mind reasons that we must logically expect *something* in return. The mind can admit that acts of true selflessness result in personal benefits, but it can't rationalize why such benefits occur. The mind can't possibly determine why we feel a sense of well-being when we have set aside something that we could have used for ourselves, to be used instead by some unknown being of some future generation. The mind may admit that acts of true stewardship results in personal satisfaction, but it can't understand why.

We practice friendship and stewardship because we *know*, deep down in our souls, these are the right things for us to do. And, we are rewarded through these simple acts because they *are* the right things to do. And we share this knowledge of these truths with all other soulful beings. This common sense of right and wrong arises from our common connection to the higher order of things, a connection we make through our soul. Common sense is a unique form of logic and rationality that arises not from the mind, but from the soul.

Humanity may well have gone as far as the human mind and body can take it without greater reliance on the human soul. In fact, our abilities to *think* and to *do* already exceed our abilities to know *how* we should think or *what* we should do. Certainly, thinking and doing will continue to be an important aspect of human evolution, but today, human values are seriously out of balance with the basic spiritual principles of right living. We are trying to control everything with our minds because we are denying, or at least ignoring, the power of our souls.

The human mind is failing in its struggle to control the "lusts" of the body, and it is utterly incapable of answering the questions of greatest importance to human civilization today. How can we resolve the conflicts within us, so we can find peace and happiness within? How can we resolve the conflicts among us, so we can live in peace and harmony with each other? How can we share the bounty of the earth with others, including those of the future, and still have enough for ourselves? How can we meet the needs of all of today and still leave equal or better opportunities for those of the future?

These questions of sustainability are not just concerning the future of humanity, but also concern the quality of our lives today. How can we add quality to our lives, rather than just quantity? How can we sustain a desirable quality of life for all people for all time? These questions and their answers are at least as much spiritual as they are mental.

Self-control is a "fruit of the spirit" – of the soul. When there is harmony and balance among the physical, mental, and spiritual, our lives will be dynamic and challenging, but they won't be in constant tension and turmoil. We won't lead lives of "quiet desperation;" instead, we will lead lives of purpose and meaning. The new era of sustainable human progress demands that we draw upon the spiritual as well as the physical and mental dimensions of our lives. The new revolution must be a spiritual revolution and the personal victory will be a spiritual victory.

The spiritual victory is within our grasp, but we should not expect it to come easy. The struggle between the body, mind, and soul is as old as human life on earth – in fact, quite likely defines the beginning of *truly human* life on earth. Every culture has its own version of the creation story, but virtually all such stories have some important commonalities. The Judeo-Christian story begins, "In the beginning, God created the heaven and the earth," and later created all of the living things upon earth. Then by one means or another, depending on the particular cultural telling the story, humans are set apart from the other living things on earth – humans are made special.

Science doesn't deny that humans somehow were created, or at least came into being, along with the other things of the earth. After all, we know that we exist. But there is no logical reason to think that humans are different, in any fundamental way, from the other living creatures. Our brains may be far more sophisticated and complex than are the brains of other species, but are not essentially different in their physical nature. So at this point in our creation stories, our minds and souls tend to lead us in two different directions. Science cannot explain the existence of the soul, so the mind is inclined to deny that it exists. However, the soul does not go away simply because the mind denies it. So the soul continues to proclaim its existence. It is quite possible we humans are special because we have a soul and because we know we have it. Without it, or without knowing we have it, we would be just another animal.

Being a Christian, although not currently a church-going Christian, the creation story I know best is the one found in the Holy Bible. The others may be just as valid, for all I know. The Bible's creation story speaks of Adam and Eve in the Garden of Eden. In Judeo-Christian theology, they were living in the garden in perfect harmony and peace with all of the rest of God's creation, quite possibly because at that time they were not fundamentally different from the rest of creation. But then, they were tempted into eating from the forbidden tree of "knowledge of good and evil." It seems quite possible that up until that time, humans, like all other creatures of the earth, could "do no evil" because they "knew no evil." God knew the difference between good and evil because God had created the tree of "knowledge of good and evil." But Adam and Eve were not satisfied being like all other creatures on earth. They wanted to be like God – they wanted to be special.

After being warned against it, the two humans in the Garden of Eden were allowed to choose to have souls, so like God, they could know the difference between good and evil. As far as we know, we were the only species that was given such a choice. The eagle that kills an innocent fish and the wolf that devours a helpless rabbit do no evil because they know no evil. But our human soul, the spiritual dimension of our being, makes us God-like. Perhaps, this is the sense in which we were "created in the image of God." However, our soul has become both our blessing and our curse. I do not believe in "original sin" or that all people are born in sin because Adam chose to eat the "forbidden fruit." My sister, Helen, believes that children are born perfect, and that all of their later faults are acquired from adults in the process of growing up. In a sense, I agree. I believe we are born as animals, with the same basic appetites and physical urges as other animals. But at some point in our development, we become aware of our uniqueness, our knowledge of good and evil. Unlike other animals, we are then capable of committing sins, whenever we allow our appetites and urges to dominate our decisions and actions. Our soul gives us a God-like responsibility of choosing between good and evil – between right and wrong and between greed and generosity. We have the knowledge of good and evil, and with that knowledge, the responsibility to reject evil in the pursuit of goodness and to reject the pursuit of narrow, short run self-interests and instead to pursue quality of life and long run sustainability.

Some scholars label stories such as those of the creation, great floods, or second comings as "cultural myths." But, such stories have been passed down from generation to generation with no known source or time of beginning. Similar stories differ among cultures, but they might do so only because they have been passed down by different people within different cultures. It is completely conceivable that similar stories within different cultures all have a common origin in actual occurrences, and thus, are not myths. These stories are called "myths" because there is no proof that the events they describe actually happened and often no logical or rational way the events could have happened, at least as told in their current versions.

However, if we suspend the tests of logic and reason, it would seem far more believable that people have chosen stories about things that actually happened, rather than simple myths, to pass down from one generation to the next. Why would people choose to pass down a "myth" as truth? Perhaps, the events didn't happen exactly as described, as the details of any story tend to change with each retelling. However, our common sense tells us that the essence of similar cultural stories, particularly the parts that they share in common, may actually be true. At some time in the past, I think it quite likely that something of the nature of the events described in these stories actually did happen.

The desire of humans to be God-like is a common theme of many cultural "myths," as well as an aspect of nearly every widely recognized, enduring religion in the world. By nature, we humans don't want to submit to some higher order, instead, we want to define or create our own order of things. We want to be our own God. However, our "knowledge of the higher order" of things, including good and evil, does not empower us to change that order. We can't make evil things good or good things evil. Our knowledge instead obligates us to make wise choices.

The age of reason, in many respects, has been an attempt to absolve humans of the responsibility of choosing between right and wrong by denying the existence of anything beyond the grasp of the human mind – beyond our ability to reason. If something couldn't be proven, it didn't exist, and thus, we didn't need to be constrained by the myth of its existence. Logic and reason would be our new God. We would define our own order of things. We would recreate the earth and the living beings upon it to fit *our* image of how they should be.

We can deny the existence of our soul, but we cannot avoid the consequences of our choices. We continue to live lives of internal conflict because we make "rational and logical" choices about how best to serve our self-interests. The world is continually confronted with conflicts and wars because we continue to make "rational and logical" choices about how best to defend ourselves against others. Starvation, disease, and disaster are everyday occurrences in many parts of the world, because people continue to make "logical and rational" choices concerning how best to survive the next day. Even the most affluent of nations are riddled with crime and drug abuse because people continue to make "logical and rational" choices concerning how to get as much as they can for themselves, and get it right now. We waste billions of dollars on courts and lawyers because we make "logical and rational" choices about how best to resolve our disputes with others. We continue to degrade the natural environment and rip apart the social fabric of our societies because we insist on making "logical and rational" choices concerning how to keep our economy growing. The age of reason may have brought us many good things in the past, but it holds no promise of a better life in the future. It's time for a spiritual revolution. We need to begin relying on a new kind of logic and rationality, the unique logic and reason of the soul.

Scientists quite simply are not credible when they go about their work as if the spiritual dimension of our being did not exist. They are not credible when they suggest that spirituality has no place in research, particularly in research that directly involves, relates to, or affects humans. They are not credible when they ignore moral truth and spiritual laws, which so affect the quality of human life. Scientists are not credible when they deny the existence of their own soul.

We don't have to convince people they have a soul to win the spiritual victory. We only have to give them the courage to admit what they already know. A spiritual awakening is already underway in American society. There is a growing awareness that the ills of our society today are a symptom of spiritual atrophy. The fundamentalist movements today are being fueled by growing disenchantment with the impacts of modern science and technology on society, regardless of whether they are Christian, Jewish, or Islamic. Problems and conflicts arise from these movements because their members are searching for new answers in religious dogma, rather trusting their own spirituality.

We know that most of today's problems are the consequences of yesterday's logical and rational solutions. We know that the mind is incapable of solving problems that are spiritual in nature. We know that it will take something other than money to alleviate poverty, something other than more courts and prisons to resolve disputes among people or to prevent crime, and something other than better weapons to stop or prevent future wars. We know that we must reawaken the spirit within us if we are to gain the insights and knowledge we need to achieve lives of quality. We know that the answers to the most important questions of our times are to be found not in our minds, but in our souls. All we need to do to win the spiritual victory is to find the courage to admit what we know to be true – to begin using our common sense.

I don't expect a great spiritual victory to be won in a decade, perhaps not even in fifty years. Maybe we will never achieve a true spiritual victory, since humans have wanted to be their own God since the beginning of time. But I do expect a spiritual revolution, or maybe a spiritual renaissance would be a better term. I also expect the quality of human life on earth to be enhanced as we reclaim the sacred in our day-to-day lives. I expect people to take better care of each other and better care of our natural environment as we accept our spiritual responsibility to do so. I expect Americans to reject endless wars of domination, whether to control the world's resources or to impose our ideology upon others. And I expect the quality of American life to improve as we become more accepting of our responsibilities toward each other and toward nature. I see no reason to accept anything less.

The spiritual renaissance will pave the way for mental victory, as we begin to rely more on our common sense of right and good to guide our thoughts and actions. Over the past several decades, we have been encouraged to think as "biological mechanisms." We have been encouraged to focus our thoughts on logical and rational means of achieving tangible and material success. The self-help gurus of the age of reason have told us how to "think and grow rich," in the corporate world, in the stock market, through individual entrepreneurship, or though the power of personal relationships. Invariably, we have been encouraged to think according to some variation to the old linear sequential industrial process. We have been exhorted to assess our opportunities, set specific goals, define measurable objectives, develop plans, devise strategies, to make decisions and act accordingly, and then to monitor progress and make strategic course corrections until we achieve our objectives. When all of our objectives are achieved, we will have achieved our goal – success. It doesn't seem to matter whether the goal is becoming a millionaire by the time you are 30 years old or marrying a rich husband or wife, the thinking processes that lead to success are much the same.

Today, however, a growing number of new self-help gurus are defining a new intellectual paradigm and are suggesting new arrangements of thoughts and actions, and are suggesting the pursuit of a more holistic, dynamic concept of success. These new ways of thinking are providing the keys to a different kind of personal victory – a victory that extends beyond individual self-interest, to include the interpersonal and spiritual layers of self. Unlike those who peddle individual, material success, they are promoting a higher overall quality of life. As with the old self-help gurus, there is a good bit of commonality among those thinking and writing about the new roads to personal success.

Be guided by purpose rather than goals. Life is a journey, not a destination. Develop and maintain a holistic vision of a desired way of living, rather than a desired level of accomplishment. A life lived with purpose and meaning is a successful life, regardless of the material accomplishment. Purpose is a far better compass than are objectives and goals.

Our preoccupation with goals and objectives, perhaps more than any other factor, account for the workaholic way of life that dominates most industrialized societies today. We are taught first by our parents, and then in classrooms, the mass media, and by example in every workplace in America: the purpose of life is to achieve something tangible. We are told we need to set lofty goals for ourselves and then set about achieving them. Achievement is the key to success.

However, setting goals almost invariably leads to attempts to control circumstances and people, who in fact are beyond our control. Goals encourage us to manipulate and abuse our relationships, and invariably detract from our overall quality of life. Instead of setting quantitative goals, we need to envision a desirable quality of life, stemming from a life of meaning and purpose. Our day-to-day decisions can then be guided by principles, to achieve a life of purpose.

Rely on principles rather than values. Principles are those things that we know to be true, right, and important, regardless of the situation. All true principles are rooted in first principles – in common sense. Values are those things we have been taught, or have learned on our own, which help us maintain consistency in our day-to-day decisions and actions. Our values may not be consistent with our principles. Never allow values to take precedent over principles.

Principles are the fundamental laws of the universe. Principles are those things that never change. Everything else in this ever-changing universe changes in relation to its unchanging principles. We typically think of principles as those expressed in physical laws – such as the laws of gravity and motion, the atomic structure of matter, or the genetic codes of living organisms. However, there is reason to believe that equally inviolate principles exist also with respect to human nature.[1]

The "natural laws" of human relationships are a part of every major and enduring religion, social philosophy, and system of ethics or morality. My own list of twenty principles for positive personal relationships is: be honest, responsible, respectful, kind, and fair; act by believing, trusting, sharing, caring, and loving; react with temperance, patience, sympathy, empathy, and forgiveness; while remaining open-minded, optimistic, hopeful, helpful, and useful. Obviously, few people can practice all of these principles consistently; certainly, I can't. But there is little doubt that strong positive relationships must be built upon such principles.

Whether we are dealing with other people or with the natural environment, we must rely on principles rather than values. Our values are based on conventional wisdom – things we have been taught, or have learned ourselves through experimentation. Principles arise from our common sense – things we know intuitively to be true.

Stay open to opportunity. The world around us is constantly changing and renewing itself. Our ability to perceive and understand the world is constantly changing, and thus, constantly renewing our minds. Our reality is dynamic, not static; it is revealed day-by-day, as we unfold the potentials before us. So we must stay awake and open to new opportunities to realize the emerging wonders of reality.

The world is always changing. Yesterday's solutions may not solve today's problems. Old ways of thinking don't allow us to seize new opportunities, or even to see them very well.[2] Reality unfolds before us in the form of changing patterns and relationships, and thus is constantly creating new wonders for our minds to explore. To explore these wonders, we have to keep our minds open, flexible, and pliable.

Learning is a life-long process, not something that we do in preparation for life. We become motivated to learn by living. We do not appreciate the value of history until we have experienced enough change within our society and begin understanding the tendency of humans to make the same mistakes, over and over. We do not appreciate the value of geography until we understand that different places have spawned different cultures and different kinds of people. We do not appreciate the value of mathematics until our minds have sufficient maturity to understand that numbers are abstractions of reality. We can never be good readers until our thirst for knowledge drives us to search for new answers in what others have written. And we cannot become good writers until we have experienced something we feel we must share with the rest of the world. Formal education does not create an educated person; at best, it only provides the initial means for a lifetime of learning. Our ways of thinking must evolve in harmony with an ever-changing world.

Build positive relationships. Human beings are social beings, by nature, and we also have a natural sense of connectedness with the earth. In fact, all things of the earth are interconnected, living and non-living. Relationships have value to people, apart from any purely individual or economic reward. We are all a part of the same whole, thus our purpose and meaning is a shared purpose and meaning. The quality of our relationships affects the quality of our lives.

People need other people. The physical things of life might seem to be most important, at least at first thought. Obviously, eating is our highest priority if we are starving. But, people have been known to lose their appetite, refuse to eat, or even kill themselves because they felt unloved, unwanted, or oppressed. It's true that a baby cannot live without milk. But, neither can a baby survive without love, at least for very long.

All things on earth are interconnected. Thus, we are related to everything else and everyone else, regardless of whether we realize it or not. When our relationships with other people are weak or negative, we feel lonely and unloved. When our relationships with others are strong and positive, we feel a sense of belonging and of affection. When our relationships with the earth, with the whole of the web of life, become weak or negative, we feel that our life has lost purpose or meaning. But when our connections to the web of life are strong and positive, we sense that what we are doing is important, that we are helping to make a better world.

Go to the source for solutions. The quick and easy ways out, often lead back in. Neutralize problems by eliminating their source. Never consider a problem solved just because you have alleviated its symptoms. Prevent future problems by maintaining healthy relationships – among people and within nature. Prevention is the only lasting solution.

Life is a complex organismic process where each action may have multiple reactions and ultimate consequences. Many of today's problems turn out to be the unintended consequences, the feedback, of yesterday's solutions.[3] In an interconnected world, we can never do "just one thing." Thus, instead of asking only "how can I stop what is happening?" we should ask also, "how can I keep this from happening?"

In living systems, cause and effect often are not closely related, with respect to either time or space. A current symptom may be the result of some minor occurrence that happened months or years ago, in some distant location, something that has gone through several cycles or iterations, accelerating or accumulating, before finally emerging as a clearly identifiable problem. In living systems, however, the biggest problems often result from the smallest causes. It may be difficult to find the real source of problems, but once we find the cause of it, and once we quit causing it, the solution may be easier than we ever thought possible.

Make good choices. Life is made up of a series of choices and consequences. We cannot reshape the world to fit our particular desires; all we can do is to choose among the available alternatives. All choices have consequences, which in turn present us with additional choices. We walk through our reality one choice at a time. The quality of our choices is the only control we have over the quality of our lives. So, learn to make good choices.

Our life is defined by the choices we make and the consequences of those choices.[4] Obviously, we can't control the world, but we can choose how we live in it. We can't control other people, but we can choose how we relate to them, and we can choose how we respond to whatever they do. We can't control the natural environment – the earth, the other living things that occupy it or the weather above it – but we can choose how we respond to it. In many cases, the choices we must make concern dealing with the consequences of our previous choices. The only control we have over the quality of our lives is the ability to make choices and to choose how we deal with the consequences. So our quality of life is a direct reflection of the quality of our choices.

Having done all, let go. Once you have confidence in your ability to respond to the challenges of life, you are free to let go and enjoy life. When you have done all you can do to ensure success, let go and let life happen.

The final key to personal victory is to learn to let go. We can't control the world and we can't control other people.[5] Thus, our need to control is a need that can never be satisfied. In addition, the need to control isolates us from the rest of the world. We can't build strong personal relationships by manipulating and controlling other people. We can't relate to the things of nature by attempting to control nature. Letting go doesn't imply that we should be controlled by others or by our circumstances, but we must remain open and vulnerable if we are to experience the fullness of life. Our security must come from our confidence that our lives have purpose and meaning, arising from a higher order of things, over which we do not and need not have control. We must learn to trust this higher order and let life happen.

I believe in these new ways of thinking about personal success because I have spent most of my life living by the old ways and have suffered the consequences. I have set and achieved goals and objectives and have found that each such success left me feeling empty and dissatisfied with life. I have accepted other peoples' values and definitions of success and found that achieving such success never left me really feeling successful. I have tried spending years learning to do things, preparing to live, only to find out that they really weren't the things that I wanted to do with my life. I have ignored the value of relationships and have reaped the rewards of loneliness and regret. I have tried the quick and easy ways of getting things done and found that they almost never stay done. I have looked back over a lifetime of choices and have regretted many of those choices I have made and their consequences. And, I have spent far too much time and energy trying to control events and people that were simply beyond my control.

I certainly cannot hold up my life, in the past or today, as an example of personal success. But I know from personal experience, to the extent that I have succeeded in incorporating these new ways of thinking into my patterns of living, day-by-day, my life has gotten better. Perhaps the most important lesson that I have learned, is that my life has a purpose and to the extent that I live in a way that fulfills that purpose in life, my life has meaning, and I am happy.

We have been told from the time when we were children that we could do anything that we really wanted to do with our lives, and we could be anything that we really wanted to be. It simply is not true! We can't do anything we want to do. Some things, for specific ones of us, are just not going to happen, no matter how much we might want them or how hard we might try. We just aren't strong enough, smart enough, fast enough, or good enough looking, no matter how much effort we put forth. But, we are strong enough, smart enough, and fast enough to do what we were meant to do.

I am convinced that each of us has a specific purpose in life. It will take all of the desire and effort that we can possibly muster just to fulfill that purpose. Some of us will be doctors, a few movie stars or sports heroes, a dozen-or-so will be presidents, but the rest of us have something more ordinary to do with our lives – although of no less importance. We should set our sights "high," because we don't know how high we are meant to go. But, we shouldn't waste our time and energy pursuing someone else's dreams, when our common sense tells us our purpose lies elsewhere. When we are fulfilling our unique purpose in life, we are living the best life we can possibly live, and in this, we will find happiness. When we have done all we can do, we must learn to let go.

Finally, our physical being is also a part of our personal being, and our personal victory will not be won unless we are being the best we can physically be. We can't all be athletes – no matter how hard we exercise or how long we train. Some of us will never even have "healthy" bodies – no matter how hard we try. We don't all have the same physical potential. But, we all have the physical ability to fulfill our purpose for being. That's all we really need to do.

Our common sense tells us that healthy bodies require good food and physical activity. If we don't feed our bodies well and use them well, they will lose their ability to function well. When the body is invaded by an infection or disease, or is injured, it may require medical treatment to assist it in healing. But a healthy body is the best means of resisting and recovering from infection, disease, or injury. You don't need a medical degree to understand the prerequisites for good physical health - it's mostly just common sense.

I have learned a good deal about diet and nutrition from my years of work on issues related to sustainable food production, although I certainly do not claim to be a nutritionist. I have learned that good food includes a variety of foods, those things for which a healthy body has a natural inclination. We may have a cultural predisposition to over-consume fats and carbohydrates, and consequently, a predisposition toward obesity. However, growing evidence indicates that obesity in America is a likely consequence of the high level of the "empty calories" in the highly processed foods being promoted by a corporately controlled food industry. The seemingly insatiable American appetite for sugar, fat, and salt may well reflect the fact that manufactured foods, although high in carbohydrates, fats, and sodium, are empty of nearly empty of all other nutrients. The body actually may be "starving" for the things that have been taken out of our foods, while it becomes obese from the things that were purposefully put in. An appetite that can be easily satisfied does not generate much profit.

The mind has the power to influence the body, just as the soul has the power to influence the mind. We have the power to reject those things that will destroy our health, and then, let our body choose things that it needs to be healthy. We can reject any dislikes we may have for specific fruits, vegetables, animal products, or grains – likely triggered by unfortunate early childhood experiences. The human body apparently is designed to be omnivorous, and thus, needs a variety of foods to be healthy. If we choose to eliminate one of the basic food groups from our diet, our body looses the ability to select a healthy diet for us, putting a far greater burden on our mind to do so.

I suffered from the common misperception that tastes in foods are somehow "hard-wired" into our brains when I first went away from home to college. I had some favorite foods but some foods I simply did not like and would not eat. In college, however, my food choices were limited. The only place I could afford to live was in the university dormitories, and everyone who lived in the dormitories at that time paid for 20 meals per week in the dormitory cafeteria, along with their rent, regardless of how many meals they choose to eat. Those dormitory meals also had very restricted choices, sometimes two meats, usually two vegetables, and maybe a salad and a couple of different desserts. I soon learned that if I didn't change my eating habits, I was going to be buying a lot of food that I wasn't eating, which I couldn't afford, and having grown up poor, was not inclined to do. So I learned to eat what was offered. And after a while, I discovered that I had developed a taste for many of the foods that I wouldn't eat a few months before. Some of my favorites today, like corn on the cob, are foods that I refused to eat as a child. I'm now convinced that I had a bad psychological experience as a child associated with corn on the cob, perhaps I ate a corn earworm with it and got sick. I am convinced also that many people are missing the enjoyment of many nutritious foods, because they are unwilling to try to develop a taste for them.

By expanding the variety in our diets, we can limit our consumption of industrial foods that are high in sugar, high in fat, high in salt, and low in nutrition. An increasing variety of locally grown, high quality, unprocessed, fresh foods are being offered all across the U.S. today. These foods may not be as quick, convenient, or cheap as are industrial foods, but being unhealthy is not convenient, medical treatment is not cheap, and recovery from illness is not quick. People can choose a common sense approach to diet and health by spending the time, effort, and money necessary to eat well.

The industrial food system is not only destructive of human health, but is among the most exploitative of all industries of its workers and of the land. Not only can people improve their personal health, but they also can help build a more sustainable society by reducing their support and eventually breaking free of their dependence on industrial foods. They can buy more of their food from local farmers markets, through subscription farming or community supported agricultual organizations (CSAs), or from local farmers who sell direct to consumers by a variety of other means. They can join food-buying cooperatives or buy from locally owned retail food markets or restaurants that buy direct from local farmers. They can buy more food from people they know and trust.

Breaking destructive habits of physical activity, like destructive diet, requires some willful influence of mind over body. Our past culture of hard physical labor has led us to place a cultural premium on physical leisure over physical activity. Thus, we must again rely on our intelligent insight, which tells us that no animal can remain physically healthy unless it remains physically active. Anyone who has been in a cast while allowing a broken bone to heal knows how quickly unused muscles can weaken and how much exercise is required to restore their strength. This same kind of atrophy occurs to all parts of the body, although more slowly and of a smaller magnitude, anytime they do not receive adequate exercise.

In general, exercise requires some level of exertion, something in addition to the effort required to breath, talk, or walk – although a "brisk" walk is among the best forms of exercise. The brain also seems to need periods of rigorous mental exercise to maintain its health and vitality. Too much exercise causes distress, rather than stress, which can be destructive. However, periodic physical exertion strengthens muscles, allowing the body to function more easily and efficiently, at rest as well as in action.

Again, when exercising, we need only rely on our common sense. If we expect to maintain a strong heart muscle, we need to exercise our heart. If we expect to have strong legs, we need to exercise our legs. If we expect to have a strong back, we need to exercise our back muscles. And, if we want a healthy mind, we have to use it now and then. We don't need to "kill ourselves" with a rigorous exercise routine, we just need to use some common sense.

We are whole beings, and thus, our personal victories must be spiritual, mental, and physical victories. All are essential to our quality of life and happiness, none is any more or less critical that the others, as was discussed at length in Chapter 7. We each have the ability to choose what kind of life we want to live – spiritually, mentally, and physically. Life inevitably involves struggles and hardships, disappointments, and heartbreaks. None of us is perfect, and we apparently need these things to nudge us along toward our intended purpose. So life inevitably involves difficulties and frustrations. But once we accept and embrace the fact that life is hard, *life is easy*.[6] Stress is necessary to build and maintain strength, if we have the wisdom to keep it from becoming distress. We didn't choose the kind of world we were born into, but we can choose how we want to live within it. We can't control what happens to us, but we can choose how we respond to what happens. Those personal choices invariably shape how we relate to other people and to our natural environment. And, as our choices begin to affect other people and the world around us, we begin little by little, to change our part of the world. That's the way change happens, one-by-one and little by little, until seemingly all at once, our little personal victories are transformed into a societal revolution.

Endnotes

[1] Stephen Covey, *Seven Habits of Highly Effective People* (New York: Simon and Schuster, 1989).

[2] Robert Kriegel, *If it ain't broke... Break it!* (New York: Warner Books, 1991).

[3] Peter Senge, *The Fifth Discipline* (New York: Doubleday Publishing Co., 1990).

[4] Alan Savory, *Holistic Resource Management* (Covelo, CA: Island Press, 1988).

[5] Susan Jeffers, *End the Struggle and Dance With Life* (New York: St Martins Press, 1997).

[6] Scott Peck, *The Road Less Traveled* (New York: Simon and Schuster, 1978).

14

RECLAIMING OUR DEMOCRACY:
THE SOCIETAL VICTORY

The great transformation will bring about fundamental change in our culture, science, society, and institutions, change perhaps more profound than ever experienced by humanity. The industrial development paradigm, with its foundation in the logic and the rationality of contemporary science, permeates all aspects of our society today. The new sustainable development paradigm, with its foundation in the insights and intuition of common sense, will permeate the whole of society in the future. In addition, the new sustainable paradigm will integrate the political, social, and economic aspects of society into a single, unified whole.

The concept of leverage will be a key element in bringing about fundamental change. A lever provides a means of increasing the power of an action by spreading its force over a greater distance, across space and over time. Organizational leverage is employed by focusing on small, doable actions or changes in specific structural components that will result in significant, enduring changes in systems as wholes. We are often frustrated in our attempts to make major, large-scale changes because we focus on trying to restructure entire organizations rather than focusing on the specific flaws in underlying structures that are creating problems for the organization as a whole. These specific flaws provide leverage points that can be used to bring about organizational change.

I learned the lesson of looking for leverage points the hard way, by experiences. During the first five years back in Missouri, I tried to work with University administrators – department heads, directors, and deans – to institute a research and education program in sustainable agriculture. I had 20-years of experience as a faculty member and administrator, and I was confident that I knew how to initiate a new program, particularly one that clearly had so much potential for the state of Missouri. Missouri is a very diverse state, with wide variety in its topography, geology, soil types, and climate. It is not well suited for the specialized, standardized, consolidated production units that characterize industrial agriculture. As a result, Missouri still has more small farmers than any other state except Texas, which is far larger. Missouri was ideally suited for the diverse, dynamic, site specific, and individualistic paradigm of sustainable farming. I couldn't see any reason the University of Missouri would not embrace a new program in sustainable agriculture.

However, I grossly underestimated the political and economic power of the industrial agricultural establishment. After five years of trying every strategy my experienced mind could conceive, I had made no significant progress in institutionalizing the sustainable agriculture program. I was coordinating an effective program, which was widely considered to be successful by my colleagues in other states. However, the program was supported almost entirely by outside grants and contracts, rather than by University funding or support. We were operating with an annual budget of three to four hundred thousand dollars at the time, but the University was only covering about half of my salary and little else. In 1994, I put together a proposal for a program in "Sustainable Grassland Farming." It seemed a natural, since Missouri also ranks second to Texas in numbers of beef cows, which for the most part, are raised on small farms on Missouri's highly productive grasslands.

Over a period of several weeks, I discussed my plan with everyone in the University and in the various state agencies that might have any interest in participating, or just being aware, of such a program. The program was to be multi-disciplinary and multi-agency, so it needed the support, or at least the acquiescence, of a lot of different administrators, as well as key faculty members. Some people seemed to have minor reservations, but no one voiced any real opposition to the proposed program. I made arrangements to announce the new program at a public event during Agricultural Science Week, a week of events attended by everyone who is anyone in Missouri agriculture.

I had been asked by various administrators to include specific faculty members from various departments on the program at the public event, I assumed to show its breadth of support. To my surprise and shock, the faculty members, one by one, focused their public remarks on their reservations and concerns, with each concluding that such a program would be a duplication of on-going programs and simply was not necessary. I had been "ambushed," publicly and effectively; the proposed program now had no chance of succeeding. I had tried to change the organization too much, too quickly, and too openly; the organization rebelled. A few weeks later, the Dean announced a virtually identical program, to be carried out within the existing academic departmental power structure – a disciplinary program with others disciplines invited to participate. Of course, this program never accomplished much because its structure was incompatible with its mission.

From the day of the public meeting forward, I concluded that I had to change my tactics completely, if I expected to accomplish anything. The fundamental flaw in existing University programs was that they were not addressing the needs of the vast majority of Missouri farmers – the small farmers for which the industrial agricultural paradigm was simply not working. But these farmers, individually, had no political power. I decided if we were to bring about institutional change, we had to find ways to reach these small farmers with programs that did work, and then, ask these farmers for their support.

So I recruited a core group of extensions agents – university faculty members who work directly with farmers and live in the rural communities where they work. By 1995, we were receiving professional development funds from USDA that were intended to familiarize all extension agents with the concepts and promises of sustainable agriculture. After five years, I had concluded that the vast majority of Missouri's extension agents already knew enough about sustainable agriculture to know that it wasn't going to do anything for them professionally, at least not in Missouri. Forcing them to set through such training would have been a waste of time and money. By concentrating our limited resources on this small group of agents, we were able to give them opportunities to do things that they truly believed in, things they would have never been able to do without us. These agents gave us the leverage we needed with a core group of small farmers, who in turn gave us leverage to begin changing the institution.

We empowered a key group of small farmers who had been ignored by their public institutions, and thus had never participated in the political process. These farmers testified at public hearings, all across the state and at the state capitol, and lobbied their local and state representatives to increase funding for sustainable agriculture in Missouri. They helped form new organizations, such as the Missouri Farmers Union, to exert political pressure for sustainable agriculture programs and policies. And they routinely confronted the agricultural establishment, including the University of Missouri, demanding that these public institutions serve the public interests.

In bringing about societal change, we often achieve too little because we attempt to do too much, or at least too much, too quickly, and too openly. We can't expect to win in a head-to-head confrontation with the politically powerful multinational corporations, but we can focus on some key flaws in organizational structure of our economy, from which corporations derive their power. We can't fix the flaws in our economy through direct confrontation, but the flaws in our economic structure could be addressed by our elected representatives in government. Our elected representatives are unwilling to address those flaws because of corporate influence on elections. However, the people still have the power to change the electoral process. Thus, the people are the point of leverage for a societal victory. To win, the people must regain control of their government by electing people to office who will represent their public interest. The final leverage point: we must reengage people in the processes of government.

Every eligible citizen has a responsibility, as well as a right, to vote. Every vote cast must be counted and every vote counted must have the potential to make a difference in the outcome of elections. Voter apathy and lack of participation reflect serious doubts regarding each of these prerequisites among many people of the United States. To solve the problems of voter apathy and skepticism we can't simply treat the symptoms; we must go to the source of the problem.

Regarding presidential elections, some have argued for abolishing the Electoral College. Many have argued for a direct vote of the people, with a uniform ballot for all national offices and a standardized procedure for casting and counting all votes in national elections. Some aspects of such proposals are worthy of serious consideration. The technologies currently used for counting votes in most counties are hopelessly out of date, and electronic voting, thus far, raises more doubts than it relieves. Requiring paper printouts of each vote in electronic voting machines is but a step in the right direction. To restore confidence in the voting process, the electronic voting network must be as secure as the current electronic financial network. We know it can be done and it would be done, if we gave our political system as high priority as we give our financial system.

However, technology will not ensure voter participation in elections. Many people don't vote because they don't feel that their vote will make any difference. A national popular vote for President would only make this matter worse. National polls taken before elections would likely identify the national vote leader going into an election. Even in cases where the race is too close to call, as in 2000, voters will know that a margin of 300,000 votes, as in 2000, would be considered a "razor thin" margin of victory in the national popular vote. What difference can one vote possibly make, when the narrowest of victories are decided by a quarter-million votes?

With the Electoral College process in place, the 2000 presidential election hinged on a reported 154-vote lead for George W. Bush in one state, Florida. Before the manual recounts were stopped, the margin was said to have dropped to close to 50 votes. Again in 2004, the presidential election hinged on a close vote in one state, that time Ohio. With national elections decided by margins this small, it's easy to see that every vote can count. The Electoral College helps to retain the importance of each vote. In addition, the Electoral College still rests on a solid foundation of the principles of a democratic republic; each state, regardless of size, has two electoral votes, one for each of its two Senators, in addition to one vote for each of its members in the House of Representatives. However, the Electoral College does need to be restructured to meet the needs of today's society.

The major problem with the current Electoral College is that nearly all states give all of their electoral votes to the candidate that wins a majority of the popular vote for the state. However, the electoral process could easily be reformed so that each state's electoral votes associated with its representation in the House would be allocated in proportion to the popular vote received by each candidate. This would allow the vote of the people to be more accurately reflected in the electoral votes cast by their state. The two remaining senatorial votes would go to the candidate who received a plurality of total popular votes for the state. This process would provide representation of the states as well as individuals in election of presidents. A presidential candidate would have an incentive to campaign for the popular vote, as well as each state's majority votes, in order to win an election.

Two states, Maine and Nebraska, currently allocate votes based on proportion of popular votes, so such a proposal is constitutional. The states have the right to decide how their electoral votes will be allocated, so such changes could be made at the state level, one-by-one. Such a change would not require a constitutional amendment nor would need to be approved by the national legislature, where corporations have their greatest power. Perhaps, such a change could fly under the corporate political radar screen, as it would have no direct effect on corporate interests. A proposal to allocate electoral votes proportional to the popular vote was recently defeated in Colorado, and the majority party in any given state will oppose anything that might give some of its current electoral votes to the opposition. But the people have the power to put the public interest ahead of politics. The people must demand change.

Under the new arrangement, any single vote could be the vote that determined whether one candidate was to receive one more, or one less, electoral vote, regardless of the size of a statewide majority. The inevitable early media forecasts of the popular vote winners of particular states would be far less significant and would be far less likely to discourage voters from going to the polls. At least 50 electoral votes, the last one to be determined in each state, would always be in doubt until each state's totals were tallied. Every vote in every state would be important, and worth campaigning for, in virtually every election.

Equally important, any candidate receiving a proportion of the popular vote equal to or greater than their proportional representation in the House would be entitled to least one electoral vote. For example, if a state has 10 members in the House of Representatives, an electoral vote would be awarded for each ten percent of the popular vote, with the last vote allocated on the basis of a plurality. Under such a process, third or fourth party candidates, who have virtually no chance of receiving electoral votes under the current winner-take-all system, would have a far better chance of affecting the outcome of elections. While the percentage threshold required to receive even one electoral vote would eliminate any frivolous candidates, the supporters of serious minor candidates would not be disenfranchised by the electoral process.

Presidential candidates would still be required to receive a majority of electoral votes to win the presidency. A candidate lacking a majority of electoral votes after the presidential election would need to negotiate with other candidates to secure sufficient votes to win the election. This would be quite similar to the way coalition governments are put together under parliamentary systems. The negotiation would simply take place at election time, giving the president sole executive power to govern. The important principle here is that supporters of third and fourth party candidates have a far better chance of influencing the outcomes of elections, even if their candidate has little chance of winning a majority. If the major candidates fail to satisfy a majority of the constituency, they know they will then have to negotiate with third or fourth party candidates in order to be elected.

Controversy concerning the Electoral College is not new and it is not going to go away. But more important, it is a point of leverage that can literally change the whole of American society. The details of election reform don't really matter all that much, but the principles reflected in the details are important. If democracy is to be restored in America, people must accept their responsibility to vote. Many will not vote unless they are assured of an equal right to vote, believe their votes will make a difference, and know their vote will be counted. Small changes in the electoral process could bring about big changes in voter participation.

Many people don't vote because they feel all elected representatives represent special interest groups, various types of corporations, rather than represent people as individuals. However, corporate influence on the political process is but a symptom of the means by which we finance elections, where political influence is bought with campaign contributions. The power to select and elect candidates to public office must be returned to the people. Only then, will elected representatives truly represent the people.

In spite of rhetoric to the contrary, it will be virtually impossible to have meaningful campaign finance reform until we reform the election process. For example, law banning "soft money" contributions – money given to political parties rather than individual candidates – simply diverted large contributions to other channels. The basic cause of campaign finance abuse is the ability of large contributors to influence the outcome of elections. As long as that ability exists, corporations will always find some means of circumventing any law that is passed.

The right to free speech is a right that should be guaranteed equally to every individual, regardless of his or her ability to buy advertising time or space in the mass media market. The Bill of Rights to the Constitution includes the freedom of speech, along with the freedoms of religion, the press, and peaceable assembly. All of these are rights to be ensured equally to all, which requires that these rights not depend upon one's economic or social status. The right to speak and to be heard on political matters was not meant to be a "commodity" for sale to the highest bidder. Speech on private matters, such as advertising products for sale, need not be so narrowly interpreted. But speech on public matters must be ensured to all equally.

Restricting our ability to influence elections to our personal ability to influence others would not be a violation of our right to free speech, but instead would be a protection of the right of free speech for all. The court of public opinion eventually must prevail over the U.S. Supreme Court on this issue. The ability to influence elections through large financial contributions makes the right to free speech unequal, giving greater powers to those with more money and denying constitutional protections to those who have less to spend. To restore integrity to the electoral process, we must restore the constitutional right to equal free speech.

Limiting political speech to individual speech would raise questions as to whether voters could have sufficient information to make informed choices among candidates. However, a very small proportion of the billions of dollars spent on political campaigns today are actually spent to inform the voters. Informing is relatively easy and inexpensive, but persuading is far more difficult and costly. It takes millions of dollars to finance campaigns today because the candidates feel they must create illusions of competency and leadership ability that simply do not exist. This is not informing the voter, this is selling the candidate.

Providing voters with the information needed to inform their voting decisions is a legitimate public service. Every voter has an equal right to be informed and every candidate should have an equal right to inform. Therefore, election campaigns should be financed by the public, not by private or public corporations or by wealthy individuals.

Every radio and television station in the country operates with a public charter, which obligates the station to allocate a certain proportion of its broadcast time to public service messages. Examples include national coverage of presidential debates and of presidential addresses to the nation. Such events are not sponsored by advertisers, they are provided as public services. Most newspapers devote large sections of their papers to reporting on public issues – particularly during periods of time leading up to elections, including side-by-side comparisons of candidates, issue by issue, prior to elections. There is nothing to prevent a nation-wide agreement between the government and the mass media to provide voters with the information they need to make informed election choices. The costs of mass media campaigns to individual candidates would then be limited to preparing information for the media and for the presentations made directly to voters.

Such campaigns could include debates, speeches on radio and television, opinion pieces in newspapers, articles in magazines, publications outlining positions of candidates on issues, handbills, even billboards and yard signs – anything that might truly inform the voter. All such activities could be overseen by public service boards or committees to ensure fairness and balance. However, anything that was accessible to one candidate would also have to be accessible to all other legitimate candidates for the same office. Corporations, unions, and special interests groups might even be allowed to help finance campaigns, but all such contributions would be put in a common pool of funds to help ensure that all have equal access to information.

Public financing of campaigns could be a critical point of leverage – a doable, feasible sequence of actions, capable of bringing about significant and lasting changes in the functioning of American government. Such a change could possibly be negotiated by a third party candidate for president who did not receive enough votes to be elected, but enough to decide which of the other two candidates would be elected. Candidates elected by publicly informed voters would owe no debts or allegiance to anyone other than the voters who elected them to office. Opening the field to all worthy candidates would greatly improve the odds of electing faithful and trustworthy representatives of the people, which is essential for our representative form of democracy.

Election and campaign finance reform will not immediately resolve the problem of corporate domination in our economy. Any leveraging process must begin with gaining a toehold. Election and campaign finance reform are doable, feasible actions that might provide a "toehold" for significant and enduring change in corporate domination of government.

Ultimately, our government must return to functioning for the public interests of the people rather than the private interests of corporations. To dismantle the current corporate welfare system, we must discredit the conventional wisdom that subsidizing corporate investments provides legitimate public benefits. We can begin by pointing out that none of the conventional assumptions linking increased corporate profits or growth with overall societal well-being is valid in today's economy, and most certainly will not be valid in the economy of the twenty-first century. Knowledge, not capital, is the factor most limiting current economic and social progress. Growth in economic output no longer translates into higher levels of employment as increasingly sophisticated information technologies are substituted for both labor and management. Fewer, rather than more, people are fully employed as a consequence of current corporate welfare.

Perhaps even greater than the government transfer of taxpayer dollars to corporations is government acquiescence to corporate exploitation of consumers in the marketplace. Economic competition is absolutely necessary, not only to ensure that suppliers provide the things that consumers want to buy but also to ensure that corporate profits are not enhanced by charging unnecessarily high prices to consumers. We may not be able to prove outright corporate collusion or price fixing, but we know that competitive markets are essential for a capitalistic economy. The corporations tout capitalistic freedoms as the philosophical foundation for their unrestrained pursuit of greed, while in fact they are destroying the very foundation of our capitalistic economy.

Corporate welfare promotes degradation of the natural environment and weakens the social fabric of our society – the very things we must have to sustain a desirable quality of life in the future. Elimination of corporate welfare is doable and feasible because it simply does not make sense for taxpayers to help finance the degradation of their own quality of life. This case can be made more effectively, once corporations lose their power to influence elections.

The corporations are not likely to give up without a fight, in spite of the fact that they no longer serve any legitimate public purpose. They may be forced to give some ground through the political process, as they did in the early 1900s. But they will find some way to fight their way back to power, unless the issue of corporatization ultimately is addressed directly through the constitutional process. The Fourteenth Amendment to the U.S. Constitution is the basis for the Court's acceptance of the premise that corporations have the same basic rights as people, "corporate personhood," although the Court has never written a formal opinion on this matter. The Fourteenth Amendment states, "All persons born or naturalized in the United States, and subject to the jurisdiction thereof, are citizens of the United States and of the State wherein they reside. No State shall make or enforce any law which shall abridge the privileges or immunities of citizens of the United States; nor shall any State deprive any person of life, liberty, or property, without due process of law; nor deny to any person within its jurisdiction the equal protection of the laws."[1] This Amendment was adopted to ensure equal protection under the law for former slaves, who had been freed by the Thirteenth Amendment. It was never intended that corporations would be included in the category of "persons born or naturalized in the United States," as several past Justices have articulated in their opposition to "corporate rights." Future Courts and legislatures might be less inclined to promote the absurd notion of corporate personhood.

Constitutional change is a major undertaking and there appears to be no groundswell of support for a constitutional amendment to eliminate corporate personhood. However, several other issues might be used to open the door to constitutional change, and once the door is open, the corporate personhood issue just might find its way inside. For example, abortion, gun control, and separation of church and state are all issues with large contingencies of voters begging for clarification through constitutional amendments.

The drafters of the Constitution clearly meant it to be a living document, capable of changing to meet changing needs over time. In the words of Thomas Jefferson, "I am not an advocate for frequent changes in laws and constitutions, but laws and institutions must go hand in hand with the progress of the human mind. As that becomes more developed, more enlightened, as new discoveries are made, new truths discovered and manners and opinions change, with the change of circumstances, institutions must advance also to keep pace with the times."[2] Thomas Paine stated, "It is perhaps impossible to establish any thing that combines principles with opinions and practice, which the progress of circumstances, through length of years, will not in some measure derange, or render inconsistent... The rights of man are the rights of all generations of men, and cannot be monopolized by any... The best constitution that could now be devised, consistent with the conditions of the present moment, may be far short of that excellence which a few years may afford."[3]

Article V of the Constitution states, "The Congress, whenever two thirds of both Houses shall deem it necessary, shall propose Amendments to this Constitution, or, on the Application of the Legislatures of two thirds of the several States, shall call a Convention for proposing Amendments, which, in either Case, shall be valid to all Intents and Purposes, as Part of this Constitution, when ratified by the Legislatures of three fourths of the several States, or by Conventions in three fourths thereof, as the one or the other Mode of Ratification may be proposed by the Congress."[4]

Some people might argue that if we are to gain the benefit of leverage, we must move slowly – changing the Constitution incrementally, one amendment as a time. However, far more leverage might be achieved by forming a coalition with all of those who want to see constitutional change to initiate a constitutional convention. Regardless of our personal positions on issues such as abortion, gun control, school prayer, or gay marriage, these issues need to be resolved through a national consensus building process. A constitutional convention could be used as a means of reaching such consensus. And perhaps more important, a constitutional convention could be called with the consent of two-thirds of the state legislatures, thus not requiring initiation from the federal level, where the corporations have their greatest power.

Such a convention would open the door to introduction of a number of amendments to ensure the long run sustainability of American society. At the top of my proposed list: All rights afforded to people of current generations shall be ensured equally to those of future generations. Such an amendment would require that every law be examined with respect to its potential impact on all future generations, and laws deemed to deprive those of future generations of their rights would be unconstitutional. Every law then would be written with explicit consideration of its future as well as present implications. Such an amendment might ultimately do more to shift American society toward ecological and social sustainability than any other single action that could be taken.

Other constitutional amendments might include the right of all people to a clean and healthy environment, the right of all people to be protected against economic exploitation, and the right of all people to adequate food, clothing, shelter, and health care to ensure opportunity for normal physical and mental development. And of course, a key amendment for sustainability would be one stating that all constitutional rights are reserved for "natural persons." All of these amendments would likely receive widespread popular support, once they were on the table for serious public discussion.

To be successful, a constitutional convention would need to open up a national dialogue concerning the fundamental principles by which all Americans agree to live. Those who are concerned that new amendments might change current judicial interpretations of issues such as the current right to have an abortion or to carry a handgun would simply have to make their case with the American people. The constitution defines the purpose and principle of a nation, and government simply cannot function effectively without such a clearly stated ethical and moral consensus of the people. A constitutional convention would also provide a powerful point of leverage to reshape the future of American society to ensure a sustainable quality of life.

I am not wise enough to resolve all of these issues, but I am wise enough to know that they eventually must be resolved. We cannot restore our democracy until we restore the basic rights of people of this generation and of all future generations to be protected from economic oppression. We cannot ensure long run sustainability until we ensure the ecological and social rights of those of future generations. The new revolution will not be over until the prerequisites for a sustainable quality of human life on earth are written into the constitutions of every industrialized nation of the world.

A change in the U.S. Constitution to ensure sustainability will require a fundamental shift in economic policy at the national level. Economic growth over the past several decades has been sustained by exploiting both natural and human capital, and such growth quite simply is not sustainable. To ensure sustainability, the government must abandon its current policies and accept its responsibility for maintaining a level of economic growth that can be sustained for the long run benefit of society.

The Federal Reserve Board provides a model that might be used to leverage changes in overall government policy. First, the Fed is an independent government agency, which is not funded directly by Congress, whose decisions regarding monetary policy do not require the approval of either Congress or the President. This allows the Fed to pursue their objectives relatively free of political pressures. This independence, as much as any single factor, accounts for the relative success of the Fed in directing monetary policy. A similarly independent Fiscal Policy Board (FPB) could be established to determine fiscal policy. The FPB would meet periodically and, in consultation with the Fed, would decide on a specific level of federal budget deficit or surplus that the government would have to maintain over specific periods of time. Congress would retain their constitutional powers to tax and spend any total amount they choose by any means they choose. They would simply have to adjust taxes and spending to achieve a budget deficit or surplus as determined by the FPB. This would allow the nation to have a reasoned approach to fiscal policy, rather than the current practice of developing unrealistic federal budgets and then dealing with whatever surplus or deficit that happens to result.

Harmonizing monetary and fiscal policy would be but the first step toward harmonizing economic, social, and environmental policies. A National Board for Harmonization (NBH) might then be formed to make decisions concerning how to balance the investment of national resources among the individual, social, and ecological economies. The NBH would not be an independent agency, because its decisions would have to be based on judgment, insight, or wisdom rather than technical expertise. They would be making decisions for the nation concerning how to balance the individual, social, and ecological economies of the nation so as to sustain a desirable quality of life for its people. The NBH would need to include representatives from both Houses of Congress and from the Executive and Judicial branches of government, and from the Fed and the new FPB. But the NBH also would need the independent representation of social, environmental, and economic interests of the public in general. The NBH would provide the broad guidelines within which the rest of government would agree to function.

The tasks of the National Board for Harmonization would not be easy, but the nation must have some means of achieving national harmony and balance among conflicting national priorities. Monetary and fiscal policy could work in harmony, if the people insisted they do so. Economic, environmental, and social policies could, and ultimately must, work in harmony, if the nation is to achieve sustainability. Checks and balances need not imply confrontation and conflict, but instead could mean a mutual commitment to balance and harmony.

Tensions and stress among the various needs of society and branches of government are both inevitable and necessary in a strong society. The purpose of harmonization is to keep positive tensions from developing into destructive conflicts and to keep stress from degenerating into distress. Somewhere within government, there must be an entity that achieves the harmony and balance that is absolutely necessary for sustaining a desirable quality of national life.

Once the economic rights of people have been assured, the personhood of corporations has been denied, and government subsidies of economic growth are eliminated, the more progressive corporations will begin to take themselves apart. Once they lose their unwarranted and indefensible power to exploit people, either as taxpayers or as consumers, they will have no other compelling reason to exist. Once they are forced to rely on productivity as their only means of sustaining profitability, they will evolve toward a more productive organizational structure, which will be far smaller, more diverse, less structured, and more decentralized in nature. As the corporations take themselves apart, stockholders may be allowed to focus their investments on smaller and smaller units of the old corporation, until in essence, they become partners with other small groups of investors, or even individual proprietors.

As the trend toward a new post-industrial economy continues, the corporate structure of business organizations will become a less significant aspect of the American economy. Thus, it will become far easier to refuse charters to those who would do nothing to promote the public good. It will also be easier to revoke the charters of those corporations that exist primarily to exploit the public. Periodic public review of corporate charters might become a potentially powerful lever for urging the economy to move more quickly and completely toward restoration of competition.

Physicists claim that one person could move the entire mass of the earth if they simply had a lever long enough and a place to stand. Such is the basic principle of leverage. If the work can be spread over sufficient time and space, even the most difficult of tasks can be accomplished with reasonable effort. Every leverage point identified in this chapter is an area where relatively little effort now could generate large eventual results. Every action or change suggested is doable and feasible and, if accomplished, could bring about significant and enduring results.

I wish I could report that we brought about dramatic changes in agricultural programs at the University of Missouri. Unfortunately, I can't. But we did make significant inroads that could never have been made without applying the concept of leverage. Moreover, the initial leverage is still working. These farmers continue to work individually and through their various organizations to exert political pressure on their public representatives, agencies and institutions, including the University of Missouri, for programs and policies that support sustainable agriculture. And over time, their numbers have grown. They have not yet won many major political victories; but by each doing what he or she can do, they have become a political force that can no longer be ignored.

Every change suggested to achieve societal victory will require support, if not action, on the part of the people in general. The individual tasks are relatively easy for each of us because they are tasks that must be shared widely among many of us. Several of the tasks that seem impossible today will become relatively easy over time. We must have the patience to begin with those that can be accomplished today; those accomplishments will give us the platforms on which to stand to leverage the changes of tomorrow. With leverage, we can move the world.

Endnotes

[1] Charters of Freedom, "Bill of Rights," <http://www.archives.gov/national-archives-experience/charters/charters.html#14>, (accessed Sept. 2006).

[2] Thomas Jefferson, "Reform of the Virginia Constitution, Letter to Samuel Kercheval, July 12, 1816," in *Thomas Jefferson Writings*, (New York Library of America Edition, Penguin Putnam, Inc., 1984), 1401.

[3] Thomas Paine, "Rights of Man," in *The Life and Major Works of Thomas Paine*, ed. Philip S. Foner (New York: The Citadel Press, 1936), 278.

[4] Charters of Freedom, "Constitution of the United States," <http://www.archives.gov/national-archives-experience/charters/charters.html#14>, (accessed September 2006).

15

CREATING A NEW WORLD OF ORDER:
THE GLOBAL VICTORY

Over the years, I have had the opportunity to travel to several other countries to speak on issues related to the sustainability of agriculture. I have had the opportunity to visit Norway, Australia, Sweden, and Korea, in addition to Canada, where I typically have made three to four trips a year. While I was still an active faculty member at the University of Missouri, I had opportunities to talk with people from a number of other countries who were visiting the University. This international experience hardly makes me an international development expert, but I am aware, from first hand experience, that other countries of the world do not necessarily share the American obsession with economic growth and accumulation of wealth.

While still at the University, I would occasionally get a call from the Dean's office asking if they could bring a visiting scholar over to my office to chat for a few minutes. Invariably, the visitor has specifically asked and insisted on talking with me because they had heard or read about our work at the University related to sustainable agriculture. I was never on the Dean's list of key faculty members to be introduced to visitors, because the University was promoting an industrial approach to international agriculture and to economic development. The University had been a strong supporter of the "green revolution," and now a prominent promoter of biotechnology, both of which are industrial approaches to farming requiring high levels of fertilizer and pesticides and large-scale irrigation systems. The Agricultural Economics Department also was an enthusiastic advocate of global free trade. Sustainable agriculture was a distraction to the University's agenda for international development programs.

When visitors from the so-called "lesser developed" countries were bought to my office, I would ask them why they were interested in visiting with me, just to get some idea of their understanding of the concept of sustainable agriculture. I would then tell them that I knew what they had been hearing in the Dean's office and from other faculty members, but that my perceptions of the challenges and opportunities of sustainable agriculture were very different from the prevailing attitudes at the University. They had been told by others, in various terms, that the industrial approach to agriculture, with its emphasis on specialization, standardization, and economies of large-scale production was the key to the economic future of "developing" nations. I suggested, however, that all countries of the world, including developing countries, have more than one logical choice for their agricultural future.

The industrial approach to development, I said, was admittedly quicker and easier than the sustainable approach. The sustainable approach would depend upon the empowerment of farmers on very modest sized farms to fit their farming enterprises, methods, and practices to their particular farms and to their particular communities. Sustainable farms are inherently dynamic, diverse, and individualistic, and thus, there is no single formula or recipe for sustainable farming. However, by fitting their farms to their unique places and unique communities, sustainable agriculture allows farms to become highly productive and profitable over time, without degrading either their land or other natural resources and without degrading the quality of life in their communities.

I would then ask if the visitors had been told that the industrial agriculture promoted to them would invariably degrade their land and destroy their communities and their rural culture in the process of increasing agricultural productivity. Every nation that successfully implemented industrial farming has experienced pollution of its streams and other sources of drinking water. Forests have been cleared and the fragile land farmed until it has been depleted of natural productivity. Smaller farmers, unable to afford costly fertilizers and pesticides cannot compete, and the land is consolidated into larger and larger units, often controlled by multinational corporations.

Subsistence farmers can no longer survive, because overproduction from large farms pushes prices so low than subsistence farmers can no longer meet even their meager cash needs with sale of their surplus production. Small farmers are forced to abandon their rural communities, in hopes of being able to somehow survive in the ghettos that invariably spring up around urban centers in industrially developing nations. Their resources are depleted, their communities are lost, and their culture is destroyed. They are left with no choices other than to embrace neoclassical capitalism or to rise up in revolution.

After explaining the alternatives to foreign visitors – the slower, more difficult approach of sustainable development or the faster, easier approach of industrial development – the visitors were no longer sure what they should think. Most clearly were not willing to sacrifice their land, their communities, and their culture for the quick and easy industrial fixes to their rural economy. They did not share the blind faith of most Americans that economic growth and accumulation of wealth are the only significant factors to be considered in economic development. I am confident if they had believed the things I was telling them was true, they would have chosen sustainable agriculture over industrial agriculture as a development strategy. They simply didn't know what to believe; I was a lonely, dissenting voice in a chorus of industrial development.

This book has focused on the challenges confronting American society and on addressing those challenges within the political, social, and economic context of the United States of America. However, the challenges are not uniquely American challenges but are global in scope. The vision for a sustainable quality of life is not just an American vision but must also be a global vision. And, the victory must ultimately be a global victory. While addressing the full dimensions of the global revolution is beyond the scope of this book, it would not be complete without giving at least some attention to the global victory.

The villain threatening the sustainability of global society today is disguised as "free trade." The threats posed by so-called global free trade are rooted in a world dominated by mechanistic technologies, industrial organizations, exploitative economics, corporatist politics, and materialistic values. All of today's critical global problems – including military conflicts, chronic malnutrition, environmental degradation, and social injustice – are symptoms of using a mechanistic worldview to address the challenges of a living world. These problems are being addressed through organizations such as the World Trade Organization (WTO), the International Monetary Fund (IMF), the World Bank (WB), the North Atlantic Treaty Organization (NATO), and the United Nations (UN).

Their obvious lack of success in solving global problems is not a reflection of the impossibility of their mission, but rather, the inadequacy of their basic worldview and operational paradigms. While some critical global problems are social, others ecological, and other political, nearly all are rooted in economics. The most pressing social, ecological, and political problems of today clearly have their roots in economic globalization.

The term, "globalization," is used most frequently in referring to "global free markets," however, the concept is far broader in meaning. To "globalize," according to Webster's dictionary, means, "to make worldwide in scope or application."[1] Individual actions and their implications become increasing global, as the actions of people in one region or nation of the world increasingly impact those of other regions and nations.

We live in a global ecosystem, regardless of whether we like it or not. We have no choice; such is the nature of "nature." The atmosphere is global. Whatever we put in the air in one place eventually may find its way to any other place on the globe. Weather is global. The warming or cooling of the oceans in one part of the world affects the weather in another, which in turn affects the temperature of oceans elsewhere on the globe. All the elements of the biosphere are interrelated and inter-connected, including its human elements. We are all members of the global community of nature. We have no choice in this matter.

Increasingly, we also live in a global "social" community. Global com-munications – print media, radio, television, and the Internet – have erased technical communications barriers among nations, resulting in the spread of common cultural values around the globe. Global travel has become faster, easier, and less expensive, resulting in greater person-to-person sharing of social and cultural values among nations. Consequent-ly, the distinctiveness of national cultures has diminished. We seem to be moving toward universal membership in a common global culture.

However, in matters of society and culture we have both the right and the responsibility to choose. We have the right to maintain whatev-er aspects of our unique local or national cultures and communities that we choose to keep. And we have the responsibility to protect this right against the economic or political forces pushing us toward a single glob-al culture or social community.

We also seem to be moving toward a single global economy. All of the national economies of the world are interconnected through their dependence upon each other for trade. Problems anywhere in the world economy can create economic problems for nations all around the globe. But in the matter of a single global market, we also have the right and the responsibility to choose. Every nation has the right to maintain those aspects of its local and national economies that it deems necessary to protect its resources and people from exploitation.

However, the so-called developed countries of the world, particular-ly the United States, are pressuring the rest of the world to accept the concept of "global free trade." This initiative is being carried through by the World Trade Organization (WTO) with financial support from the International Monetary Fund and the World Bank. The multinational cor-porate community, clearly the motivating force behind the initiative, is already able to move capital, technology, and goods and services among countries in which they operate through intra-firm transfers. However, they want to have free rein to treat the world as a single global market – without quotas, tariffs, or any form of restriction. They want free and unfettered access to all consumers, workers, and natural resources.

Efforts of the WTO to establish global free trade have encountered strong resistance, not only from those in the lesser-developed nations, but also, from many groups of people within the developed nations. Increasingly, people are beginning to realize that a global free market eventually would destroy the social and political boundaries that now protect nations from economic exploitation by the multinational corporations. The crux of the WTO controversy is not whether we are to have a global economy – we already have one – but instead, whether all national economic boundaries will be eliminated, resulting in a single global free market.

Perhaps the best way to understand the implications of market globalization is to begin by addressing the issue of boundaries, in general, and to ask why boundaries now exist in nature, in society, and in economics. The boundaries in nature, the ecological boundaries, were put there by natural processes. Such physical features as oceans, mountains, and even rivers and ridges, separate one physical bioregion from another. Such boundaries define the inherent diversity of healthy natural ecosystems. Boundaries separate and define the form or structure of those things that support life: sunlight, air, water, and soil. Boundaries also define the physical structure of all living things: bacteria, fungi, plants, animals, and humans. We know also that biological diversity is necessary for life; diversity that distinguishes cells, organs, and living organisms from each other; diversity that gives resistance, resilience, and the regeneration ability to living communities. Without diversity, without boundaries, nature could not sustain life, including human life.

Cultural and political boundaries are those things that define distinct "communities" of people – including cities, states, and nations. Societies have established such boundaries to facilitate relationships among people within boundaries and to differentiate between relationships among people within a given community and their relationships with people in other communities. Within cultural boundaries, relationships were nurtured to enhance social connectedness and personal security. Boundaries between communities maintain some sense of identity, and thus, maintain diversity among different groups or collections of people. Diversity among communities maintains choices and opportunity both for those of the current generation and for those of generations to follow. Historically, whenever one human culture or society has become dominate, but then later has failed, alternative cultures and societies have always been available to restore health and growth, and thus, to provide resilience, and long run security for human progress. Without cultural diversity, there would have been nothing to replace the long line of failed societies of the past.

The basic purpose of economic boundaries is to promote free trade within the boundaries of communities and to carry out selective trade among those communities that are separated by economic boundaries. Economic diversity, as defined by economic boundaries, is necessary for division of labor and specialization. If all national economies were to lose their distinctiveness, all nations becoming as one, all potential gains from trade among nations would disappear. Historically, economic diversity among nations also has been considered a necessity to ensure choice and opportunity – to ensure health, growth, resilience, and long run security of the global economy. Humanity has not been willing to put "all of its economic eggs in one basket."

Today, however, the major economic powers of the world are pressuring the rest of the world to put all their "economic eggs" in the "WTO basket." These leaders are motivated more by short run economic consideration than by longer run concerns for human culture or for the natural environment. In this respect, other nations quite likely are being misled by the "economic culture" of the U.S., which now dominates the global economy. The tremendous growth of the U.S. economy over the past century is widely attributed to our so-called free market economy. Because of this growth, a new culture of economics now holds sway among many of the most economically powerful nations of the world.

Within this culture, economic boundaries are viewed as obstacles to trade, as limiting the ability of investors to maximize economic efficiency. Free trade among all nations would result in a more efficient global economy, they say, thus benefiting all people of the world. Current barriers to trade are nothing more than artificial, political restraints designed to protect specific individuals and industries, within nations, against economic competition from more efficient producers, in other nations. Thus, the WTO should work to remove such barriers, allowing the most efficient producers in the world to produce the world's goods and services, resulting in the lowest possible cost of goods and services to consumers everywhere – so they claim.

Such claims are based on theories of free trade that have become something of sacred tenets of economics. These theories had their foundation in the early 1800s, primarily in the writing of British economist, David Ricardo.[2] Ricardo showed that when two individuals choose to trade, each is better off after the trade than before. People have different tastes and preferences, and thus, each person values the same things somewhat differently. So, if I value something you own more highly than I value something I own, and you value the thing that I own more highly than you value the thing you own, we will both gain by trading. I will get something that I value more than the thing I now own and so will you.

The same concept can be used to show the potential gains from specialization. Even if one person is not as efficient in doing much of anything as another, one will always have a "comparative advantage" in specializing in what they do "relatively more efficiently" and trading their surplus production to the other.[3] For example, people in lesser-developed countries tend to have a "comparative advantage" in low-skilled, hand labor, not because they are necessarily more efficient in those tasks, but because they have few other employment opportunities. Specialization and trade allows them to put their skills to the highest valued uses available, even if their opportunities are few.

The problems of free trade arise because trade between two independent individuals, in the context of the early 1800s, does not accurately reflect the reality of trade among nations in the early 2000s. First, trade is truly "free," in an economic sense, only if both partners are "free not to trade." Participants in free trade must have an interdependent relationship. Interdependence implies that people relate to each other by choice, not by necessity. If one trading partner is dependent on another, the dependent partner may have no choice but to do whatever is necessary to maintain the relationship. Dependent trade is not free trade.

Poor nations are made dependent on rich nations by their lack of economic wealth, economic infrastructure, and technological advantage, regardless of their inherent worth to humanity. In many cases, rich nations exploit the workers and resources of poor nations through trade, because the poor nation sees no other way to avoid physical depravation or starvation of their people. In some cases, nations feel compelled to trade with other nations with whom they have national defense agreements, because they have no other means of defending themselves. Trade when one party feels compelled to trade is not free trade.

Second, free trade must be informed trade. Both parties must understand the ultimate consequences of their actions. When a developed nation encourages a lesser-developed nation to produce for export markets, knowing that such production will lead to exploitation of their natural and human resources, and does not inform them of the consequences, this is not free trade. The leaders of the lesser-developed nation may reap benefits from such trades, often including bribes or payoffs from the outside exploiters, but the resources of the lesser-developed nation will be exploited rather than developed. The people will be left with fewer opportunities for developing their country than they had before. The exploiters know the consequences but the exploited do not. Uninformed trade is not free trade.

Third, free trade, in economic theory, implies that the decision is made by an individual, not a collection of people, or a nation. Individuals are whole people, presumably absent of unresolved internal conflicts regarding the relative values of items to be traded. A person trades only if they decide trading is good for them as a whole. Nations cannot think with one mind or speak with one voice. Nations, as large collections of individuals, may make and carry out trade agreements to which a substantial portion of the nation's population is opposed, both before and after trade takes place. In economics, it doesn't really matter how many people are made relatively worse or better off by trade, as long as trade results in greater total wealth and growth of the overall economy. Free trade doesn't address issues of social equity or justice.

Fourth, the principles of economic trade theory are rooted in a barter economy where people trade things rather than buy or sell things. In an international currency economy, comparative advantages in trade can be distorted by fluctuations in currency exchange rates that have nothing to do with relative productivity. Such fluctuations can cause the exports from one nation to become more or less costly to importers from another nation for reasons totally unrelated to changes in production efficiency. Under such conditions, free markets do not result in efficient resource use.

Finally, in classic trade theory, each trading partner uses their individual resources, land, labor, capital, and technology to do whatever they do best – to realize their comparative advantage. No consideration is given to the possibility that one nation might instead transfer some of their resources, such as capital and production technology, to another nation where they might generate even greater profits. Mobility of capital and technology, hallmarks of today's global economy, eliminates the comparative advantage of higher cost nations, forcing them to import from lower cost nations, devaluing both land and labor in the higher cost nation to globally competitive levels. The classical economic concepts of comparative advantage did not anticipate international mobility of capital and technology.

Because of all these inconsistencies between economic theory and economic reality, the theory of economic free trade does not reflect the reality of international free trade in today's world. Perhaps more important, opposition and open defiance of the WTO, from countries all around the globe, indicates that the expansion of trade being forced upon unwilling people by the WTO, almost certainly will not be true free trade.

The globalization of markets has critically important implications for long run sustainability. The boundaries in nature define the diversity of landscapes, life forms, and resources needed to support healthy, natural, sustainable production processes. Fencerows, forest edges, streams, and ridges define unique watersheds, ecosystems, and bioregions within which nature can sustain different types of human enterprises. Globalization would remove the fencerows and forests, divert the streams, and level the ridges, to facilitate standardization and homogenization of production processes. The natural boundaries needed for sustainability will be removed to achieve greater economic efficiency. Globalization of markets ultimately will lead to the loss of ecological sustainability.

Socially sustainable systems must function in harmony with human communities – including towns, cities, and nations. Social and cultural boundaries define those communities. Humanity is inherently diverse. Diversity among people is necessary for interdependent relationships – relationships of choice among unique, independent individuals. Although we have our humanity in common, each person is unique, and we need unique human communities within which to express our uniqueness. Globalization will remove the social and cultural boundaries to achieve a homogenous global society. The natural boundaries needed to sustain social responsibility will be removed to achieve greater economic efficiency. Globalization of markets ultimately will lead to the loss of social sustainability.

Economically sustainable systems must facilitate harmonious relationships among people and between people and their natural environment. The inherent diversity of nature and of humanity must be reflected in diversity of the economy. Although potential gains from specialization are real, such gains are based on the premise that people and resources are inherently diverse, with unique abilities to contribute to the economy. Classical competitive capitalism is based on the premise that individual entrepreneurs make individual decisions and accept individual responsibility for their actions. If globalization is allowed to destroy the boundaries that define the diversity of nature and people, then it will destroy both the efficiency and sustainability of the economy. Globalization of markets ultimately will lead to the loss of economic sustainability.

The new vision of a sustainable global society is a society that functions as a dynamic, living system, continually renewing and regenerating its natural and social resources. Living things are inherently diverse, and their diversity is defined by special kinds of boundaries. I recall learning about the special characteristics of living boundaries in a high school science class – they are called "semi-permeable." The walls of living cells, for example, let some things pass through but keep other things in and out – so they are called semi-permeable. If the cells in our body either were permeable or impermeable, rather than semi-permeable, they would not support life. If they didn't keep anything in, we would dry up. If they didn't let anything out, we would blow up. If they weren't semi-permeable, they wouldn't be able to retain moisture or minerals; they wouldn't be able to metabolize food, release energy, or eliminate waste. We would die. All living things are made up of cells defined by semi-permeable boundaries.

This principle of living boundaries extends to all other aspects of life. Living organisms are defined by boundaries – skin, bark, leaf surface, scales, etc. – which give them form and identity. As with cells, the boundaries of organisms must be semi-permeable or selective with respect to what they allow to pass through and what they keep or allow in and out. Without semi-permeable boundaries, life cannot exist.

Families, communities, states, and nations are all defined by social or political boundaries, which distinguish between those who belong and those who do not. Again, living social and cultural boundaries are semi-permeable or selective, by design and by necessity. Without personal, cultural, and political boundaries, human civilization, as we know it, could not exist. Without civilized human behavior, life on earth might well cease to exist. Selective boundaries are necessary for life.

To win the global victory of sustainability and quality of life, we must find ways to maintain the healthy, semi-permeable ecological, social, and economic boundaries. The answer will not be found in isolation or self-sufficiency. Ecological, social, and economic relationships among nations are essential to sustainability. Neither will the answer be found in the dissolution of all boundaries. Ecological, social, and economic identities are essential for sustainability. The key will be to find harmony and balance between vulnerability and security in national relationships – as in positive personal relationships. People must learn to make common sense decisions concerning what they allow to move across ecological, social, and economic boundaries, and what they choose to keep inside those boundaries. People must demand the right and accept the responsibility for deciding how relationships within boundaries need to be different from relationships across boundaries in order to sustain a desirable quality of life.

As with the personal and social victories, we already have the basic organizational mechanisms in place to ensure a global victory. We need only find the courage to use them. The World Trade Organization can be redirected to ensure that every nation of the world has the right and accepts the responsibility to protect its people and its resources from exploitation. The U.S. Congress has the constitutional authority to make foreign trade policy, and Congress cannot be allowed to turn its constitutional responsibilities over to a group of economic technocrats. The WTO, which was established to remove boundaries, can be redirected to ensure that boundaries remain permeable – but only to the extent necessary to ensure the ecological, social, and economic integrity of every nation. The new WTO can be used to ensure that nations neither close their borders to relationships with the rest of the world nor open their borders to outside ecological or social exploitation for short run economic gain.

On broader issues, the United Nations could be redesigned and redirected to become not only the arbitrator of political and ecological relationships among countries, but of economic relationships as well. The WTO, IMF, and WB should all be brought under the auspices of the UN, to ensure harmony and balance among the ecological, social, and economic development within the world community. The global victory will be won not through a "new world order," but through a "new world of order." In an increasingly interconnected world, a means must be found for defining acceptable and unacceptable relationships among nations. An international rule of law must be established to replace the international rule of might. Each nation should be ensured of the sovereignty it needs to maintain its ecological, cultural, and economic identity and integrity, but must be willing to engage in a process of international consensus concerning how it will relate to other nations. A global consensus must arise out of this process concerning the responsibility of each nation to develop its resources in ways that ensure the ecological, social, and economic sustainability of the global community as a whole.

My work in sustainability has led me to become personally involved in a number of international issues. I have participated in public forums, written letters to editors and opinion pieces for newspapers, and marched in the streets to try to make points that I feel are critical to our future. I have opposed the out-sourcing of America's middle-class jobs, not because I want to deny economic opportunities to other countries, but because the global race to the bottom results in the exploitation of workers everywhere. The proclaimed benefits in terms of lower costs to consumers ignore the inherent social and ecological costs of exploitative economic development and international trade relationships. I do not oppose free trade, but insist that the people of each nation have an inherent right and responsibility to choose not to trade, if the proposed trade is not clearly mutually beneficial.

I have opposed using military might as a means of gaining and ensuring continued access to the natural and human resources of other nations in order to sustain economic growth. Military spending only serves to mask the underlying weakness in an exploitative nation that cannot maintain its productivity without massive deficit government spending. Deficit spending to build productive infrastructure or to enhance the productive capacity of people is fundamentally different from deficit spending to support consumptive spending, and nothing is more consumptive than war. The former creates future benefits; the lat-

ter creates future enemies. Military force can never secure the economic foundation of a socially and morally sustainable society.

The people who fear a new world order of peace and sustainability
fear a world in which nations will have lost their national sovereignty to
some form of one world government. The new world of order, however,
will ensure the sovereignty of each nation, by ensuring its right to protect its people and resources from outside exploitation. The new sustainable world of order will ensure that all nations can and will do their part
to ensure the long run sustainability of global human society.

Perhaps the establishment of this new world of order will require a
restructuring of the UN to make it more representative of global interests. Perhaps the Security Council needs to be expanded, with permanent members representing all of the most populous nations or regions
of the world, and maybe the general assembly, which represents all
nations, needs more authority. Perhaps the WTO, IMF, and WB need to be
restructured into a single organization before they are brought under UN
authority. Such details are beyond the scope of this book. While such
actions inevitably would require long and tedious negotiations among
nations, this has always been the nature of global processes carried out
in building a better world. More importantly, all of these things are
doable.

All of the basic international mechanisms are in place to help create
a new sustainable world of order, once the global revolution has begun.
But, these international organizations are currently under the strong
influence, if not outright control, of the multinational corporations. The
international organizations – WTO, IMF, WB, UN, etc. – were designed by
people, approved by people, and joined by nations of people, for the
expressed purpose of serving the needs of the people of the world. It's
time for the people of the world to take these organizations back from
the corporations and the global politicians that serve the corporations.
All that is lacking to ensure a global victory is for the people of the world
to find the courage to demand their basic human rights – to revolt.

Endnotes

[1] *Merriam-Webster's Collegiate Dictionary*, 11th edition, "globalize" (Springfield, MA: Merriam-Webster Inc., 2003).

[2] David Ricardo, *The Works and Correspondence of David Ricardo*, ed. Piero Sraffa (Cambridge: British Royal Society, 1951-55), 132.

[3] Ricardo, *Works*, 134-136.

16

SOWING SEEDS OF HOPE

Do I actually believe I can change the world? You bet I do! I can, and
so can you. We are in the midst of a great transformation. The world is
changing as we move into the new post-industrial era, and I am helping
to determine what it will be like in the future. We all are. Modern science
has shown us how to do things, but has given us no sense of why we
should or shouldn't do them. Without a clear sense of purpose, we are
systematically destroying human civilization and we are destroying our
natural environment. Our current society is not sustainable. It must be
changed. If you and I don't change it for the better, who will?

The world will not be changed for the better by those in positions
of economic or political power. The powerful have achieved their positions of influence within the context of the world as it is. They don't
want to change the world, at least not in any way that might threaten
their positions of power. If we start questioning why we are doing what
we are doing, we might demand change. Change is inherently most
threatening to the most powerful. They have the most to lose. The power
to change rests with those of us who might appear to be powerless. We
are the people least threatened by change because we have the least to
lose.

The powerful stay powerful only by maintaining an illusion that the
weak will suffer most from change. They keep the rest of us economically and psychologically vulnerable. They instill a fear of change among
those of us who have the most to gain and rely on fear to cloud our common sense of a need for change. They want us to be afraid to ask why.
But, we – the common, ordinary people – have the power to change the
world. We are the only people with the power to change the world. But
in this time of transformation, we must find the courage to confront our
fears and to rely on our intelligent insight – our own common sense.

Lasting societal change must arise from within the people. We will
not be able to enforce laws to protect the natural environment from

203

exploitation, even if we could pass them, until the people feel a strong sense of stewardship. We couldn't enforce laws to protect people from exploitation by other people, even if we had them, until the people share a strong sense of compassion and caring. We can't legislate changes in conscience. Laws can help create awareness of problems and can demonstrate the nature of change, and exposure can sometimes change conscience. Ultimately, however, people find ways to circumvent laws that don't reflect a public consensus. The most that changes in laws can do is enforce the consensus of the many upon those few who refuse to conform or participate in processes of civilized society.

A change in the Constitution, can't create a new national consensus, it can only document that a new consensus has been reached. The new revolution will be won, not because we have changed the Constitution, but because the people have reached a consensus that allowed the Constitution to be changed. Changes in laws, rules, constitutions, and other formal codes of behavior will be lasting only if they reflect changes in the hearts and minds of the people. Lasting change must arise from within. Constitutional change must reflect a change in ideals – in principles, ethics, and standards of success.

To change the world, we must start by changing ourselves. What we do arises from who we are; we are human beings, not human doings. The essence of our being is in the purpose and meaning of our lives, which we can discover through our intelligent insight or common sense. In discovering who we really are, we may find that what we are now doing is inconsistent with our being. If so, we won't be comfortable until we change what we do to make it conform to who we are. But most important, when we discover who we are, and then change what we do, we will have begun to change the world for the better.

What difference can a single person, a few people, or even quite a few people make? There are so many people – in the world, in most nations, and even in many communities – what difference can we possibly make? These are logical, rational questions, but logic and rationality do not necessarily make them important concerns. Margaret Mead, a noted 20th century American anthropologist once wrote, "Never doubt that a small group of thoughtful, committed citizens can change the world. Indeed, it's the only thing that ever has."[1] She wasn't talking about a few powerful people forcing change upon others. She was talking about the small groups of thoughtful, committed people who change their own lives, and by so doing, set in motion chain reactions of influence, that ultimately change the world.

Every major social, political, or religious movement has begun with a small group of thoughtful, committed people who believed they could change the world. Christianity began with a young itinerant Jewish preacher and twelve disciples from very ordinary backgrounds. After seven years of recruiting, Mohammed is thought to have gained only about a hundred Islamic followers. The modern concept of Democracy began in ancient Greece and Rome where the self-governing groups were so small that each person could speak and could vote individually on each issue. The American Revolution was initiated by a handful of radicals in the Continental Congress. The radicals were unable to mount a majority until one man, Thomas Paine, wrote the pamphlet, *Common Sense*, and asked whether "a continent should continue to be ruled by an Island."[2] Another, Patrick Henry, rose and said, "I know not what course others may take, but as for me: Give me liberty, or give me death!"[3] One man joined another in the support of a few, their vision spread, and the world was forever changed.

However, it takes courage to start a revolution. The powerful defenders of the status quo may become ruthless when challenged. Confrontation, conflicts, frustrations, and failures often mark the road to ultimate success. Most of the founding fathers of America, those who pledged "their lives, their fortune, and their sacred honor" to the cause, eventually lost their fortunes and some lost their lives, but they retained their honor and they won the revolution. In the words of Susan B. Anthony, a leader of the women's suffrage movement in the U.S., "Cautious, careful people, always casting about to preserve their reputation and social standing, never can bring about reform. Those who are really in earnest must be willing to be anything or nothing in the world's estimation."[4] One person can help change the world for the better, but only if they have the courage to stand by their conviction.

Now is the time for a new revolution. Humanity is moving toward a fundamentally different era in its history. All we have to do to help shape the new world is to engage our hearts and minds in the revolutionary process.

All things are interconnected. We can't do anything without affecting everything else. Every discussion that takes place between two individuals affects, in some small way, every future discussion that takes place between any other two individuals. Every position you or I take on an issue, regardless of how strongly or widely we defend it, affects, in some small way, the positions taken by everyone else on every other issue. Every vote that we cast, every letter we write to the editor, every time we join a political campaign, or run for any office, everything we do helps to shape the attitudes, opinions, and actions of everyone else.

We are an important element of this universe. What we do really does matter. We are not separate parts of some collective that we call the world, but instead, we each are an integral part of a single whole that is the world. If one of us passes away from the earth, we do not still have the same collective, minus one person; instead, we have a new and different whole, indeed, a new world. Everything each of us does affects everyone and everything else on the earth. It's true that any individual action may not change the world very much, at least not right now. But eventually, that one action may help change the world a whole lot. We simply don't know, we can't know for sure, just how important each of our actions might be. We do know for sure that everything of any significance that has happened to humanity in the past began with the thoughts and actions of one person, or at most, a "small group of thoughtful, committed people." All we have to do to build a better society is to begin thinking and acting as responsible members of a better society, and a better society will emerge from our thoughts and actions.

As I was thinking about how individual actions can change the world, I came across a book by Malcolm Gladwell, *The Tipping Point*.[5] In the book, Gladwell likens the changes that take place within society to outbreaks of disease epidemics. In spite of his organismic analogy, he spends most of the book analyzing the process of social change, as if he were dealing with a mechanistic process. His mechanistic analogies detracts only somewhat from the otherwise powerful concept that changes in living systems occur by processes that do not conform to mathematical or mechanical rules. Epidemics do not spread exponentially, as with compound interest. Epidemics spread in ways that cannot be expressed mathematically – in ways unique to living systems.

Gladwell's three rules of epidemics include the "law of the few," the "stickiness factor," and the "power of context." The "law of the few" suggests that some few people, very specific types of people, are far more effective than are others in starting an epidemic of change. George Washington was such a person; his life touched the lives of so many. One key to changing the world is to get a few such people on your side.

The "stickiness factor" is a lot like the principle of leverage. Some ideas just seem to "stick" while others don't, and most important, the difference between those that do and don't are oftentimes very small and seemingly insignificant. So, one key to changing the world is to keep tinkering with common sense ideas until you find a "sticky" way of saying them – a way that rings true. Thomas Paine was such a person. His ideas "stuck" and his words inspired people to action.

Finally, the "power of context" suggests that ultimately whether or not an idea becomes an epidemic depends not so much on who spreads it, or its basic appeal, but whether or not the world is ready for it. The efforts of George Washington and Thomas Paine would have been all for naught, if the people of the American colonies had not been ready for revolution. The good news is, in a world that is ready for revolution, today we don't need a George Washington to lead it or a Thomas Paine to support it. All we need is a small group of thoughtful, committed people who realize the world is ready to be changed.

Tocqueville, in the early 1800s, suggested that revolutions would be far less frequent in democracies than in oligarchies because in democracies people are more nearly equal, both in wealth and in opportunity.[6] When everyone is sharing in the bounty, everyone has something to lose if anything disrupts the status quo. Also, when everyone has a realistic opportunity to join the ranks of the wealthy and influential, it is difficult to mount a revolution against wealth and influence. He also thought that a revolution would be more difficult to initiate within democracies because a strong democratic government represents the will of the "majority," rather than the rule of the "privileged few."

However, Tocqueville believed that the rule of the majority eventually might become far more powerful, and tyrannical, than the rule of any monarch. It's far more difficult to oppose the majority of the people than to oppose a single dictator. And, as indicated before, he warned that a corporate, industrial oligarchy might reemerge if Americans failed to remain vigilant.

Tocqueville's fears were well founded. A corporate oligarchy quite clearly has emerged and has fundamentally transformed our American democracy. Our democracy no longer assures equity in either wealth or opportunity. The rich are getting richer, the poor are getting poorer, and there is no serious attempt by government even to address issues of economic equity. Economic opportunities are becoming even more inequitable than is wealth, and there is no serious effort by government to address issues of social justice. The power of our central government no longer reflects the will of the broad majority, but instead the will of the narrow, corporate oligarchy. People are reluctant to rebel, because they are intentionally misled to believe that they are but a small, disgruntled minority in a functioning democracy. They are intimidated by the illusion of majority rule.

Growing inequities in wealth and opportunity, coupled with corporate domination of government, are creating a national context that is ripe for revolutionary change. Today, far more people have far less to lose and far less hope of joining the ranks of the economic and political elite, making radical change to them both less threatening and more urgent, even Tocqueville might well agree.

The "tipping point" in an epidemic occurs at the point when the spread of a disease not only accelerates, but actually accelerates in the rate at which it is spreading. The epidemic literally explodes. At the tipping point, each new case not only results in some multiple or geometric expansion of new cases, but also adds to the size of the expansion factor. Each new case brings in new victims from the ranks of "the few" who are extremely effective in spreading diseases. Each new case increases the "stickiness" of the disease as it adapts to the weaknesses of its victims. Each new case creates a more hospitable "context" or environment for the next, as it weakens the resistance and increases the vulnerability of those who must now care for the sick. Before the disease reaches the tipping point, it is present but still manageable. Beyond the tipping point, the spread becomes uncontrollable - it is then destined to run its course.

Epidemics most certainly are not limited to diseases or physical ailments. The Great Depression of the 1930s was a classic example of an economic epidemic running rampant through a sick economy. A wildfire becomes an epidemic when its towering flames reach a point where they alter the physical climate to accommodate its rage and drive those who would fight it into retreat. Nor are epidemics limited to physical or social maladies. Society can experience uncontrollable outbreaks of wellness and goodness, situations where good things accelerate the rates of increasing health and happiness.

A rapid economic expansion is a prime example of an economic epidemic. At some point, each bit of positive economic news begins to accelerate the rate of expansion. Without intervention by the Fed, in the form of hikes in interest rates, the economy might well continue spiraling out of control upwardly until it peaks and collapses. Such an expansion is not a mechanistic, mathematical economic expansion - it is an epidemic. So, epidemics occur in social, economic, and physical phenomena, and equally important, epidemics can break out in positive as well as negative directions.

American society is ripe today for an "epidemic of ideals." We could see an outbreak of common sense at any time, sweeping across this country like a wildfire. Most people in this country are weary of the social and ecological exploitation by the industrial, corporatist society. Their intelligent insight tells them that something is very wrong. They just don't yet realize the true source of their tired and run-down feeling.

The social context is right for such an epidemic. Everyone possesses a genetic vulnerability to this germ, because all have common sense, regardless of how long it has been since we used it. This country has always had a "low-grade infection" of common sense. Some people have never quit using it. Recently, however, an increasing number of people are going through personal revolutions and are returning to their common sense as a means of putting purpose and meaning back into their lives.

In their book *The Cultural Creatives*, Paul Ray and Sherry Anderson claim, "50 million [Americans] are changing the world."[7] These people believe relationships are very important, share a strong sense of community, are committed to social equity and justice, believe that nature is sacred, and are concerned for the natural environment and ecological sustainability. They also tend to be more altruistic, idealistic, optimistic, and spiritual than is the average American. They also are less materialistic; less concerned about job prospects, and have fewer financial concerns than do most Americans. These are characteristics of people who are using their common sense.

These cultural creatives are the carriers of the new epidemic of societal change. They have joined together in various social movements, including those advocating social justice, civil rights, human rights, world peace, environmental protection, sustainable development, holistic health, organic foods, and spiritual psychology. These common sense issues are merging into a global movement committed to building a more sustainable human society. Although the cultural creatives remain a minority, these carriers of the "infection" of change are growing in numbers to a point where an epidemic of change could break out at any time. That is why each person is so important. At some unknown point, the common sense infection will reach its tipping point, and then, the action of one additional person will cause it to explode into a full-blown epidemic.

At the tipping point, more people of influence will begin to influence still more people of influence with the common sense message of the necessity of building a more sustainable society for the future and a more desirable quality of life today. I recently heard Paul Hawken, author of the popular book, *Ecology of Commerce*, say that he has stopped talking so much about ecological and environmental issues and has started talking a lot more about quality of life. He wants to focus on the higher quality of life that arises from ecologically and socially responsible living.[8] We are all searching for "sticky" ways of sharing the concept of sustainability, so people can understand the message; and we are beginning to find those ways.

But most important, the tipping point will come when enough people find the courage to challenge the conventional wisdom of logic and rationality and to rely instead on the wisdom of their own common sense. The tipping point will come when enough people realize that their lives are not made better by having more stuff, but instead by caring more for other people and by leaving something more for the next

generation. The tipping point will come when enough people understand the fallacy of the promises of the economic self-interest and begin
striving to balance the economic, social, and ecological dimensions of
their lives. As each of us changes our little part of the world, we change
world context, and we move the world closer to the common sense tipping point. We can't all be opinion shapers, we can't all be wordsmiths,
but we can all change our little piece of the world. That's all we need to
do.

How can I be sure that the world is nearing a time of revolutionary
change? I can't be sure, but I can still have hope. Hope is grounded in
the possibility of something better, not the certainty of something better.
But, we only need hope, not certainty, to motivate us to continue to work
toward a better world, for ourselves and for others, both now and in the
future. I have hope that we are nearing a time of great transformation in
human history that quite literally will change the world. And hope is all
I need.

In the words of Vaclav Havel, the writer, reformer, and first President
of the Czech Republic:

"Hope is not the same as joy when things are going well, or the willingness to invest in enterprises that are obviously headed for early success, but rather an ability to work for something to succeed. Hope is def-
initely not the same thing as optimism. It's not the conviction that something will turn out well, but the certainty that something makes sense,
regardless of how it turns out. It is this hope, above all, that gives us
strength to live and to continually try new things, even in conditions that
seem hopeless."[9]

When we look around us, we might see little evidence of a new revolution. People seem to be too busy working and shopping to pay much
attention to other people or to the natural environment. Empathy and
ethics may seem to be losing ground to selfishness and greed. Things are
not necessarily going well and there may be little reason for joy. But as
long as we have the ability to continue believing in and working for
something better, there is hope.

More and more people are becoming disillusioned with materialism
as the guiding principle for their lives and for society in general. More
and more people are turning to friends and family to fill the emptiness
that can never be filled by more stuff. More and more people are turning to spirituality in search of purpose and meaning for their lives. More
and more people are learning that a life of quality is a life of balance and
harmony. These thoughtful people have discovered new principles, a
new ethic, and have chosen new standards of success. Their numbers
may still be few and there may be little indication of early success, but

these people have the ability to continue working for success, no matter what the odds. In their commitment, there is hope.

There might seem to be little reason for optimism. The corporate interests are so powerful – politically and economically – they seem to control everything. How can the people ever again expect to prevail against such odds? But there have been earlier times in America, as well as in other parts of the world, when it seemed that the oppression of the powerful over the weak was destined to continue forever. But people have always been able to overcome oppression, regardless of the odds. The so-called political experts that the people of the old Soviet Union could never expect to gain their freedom, the power of the communist state was simply too strong. Most people of the Soviet Union apparently shared this gloomy assessment. But, hope does not depend on our conviction that we will eventually prevail. We need only be convinced that what we are doing makes common sense, regardless of how things eventually turn out. When we live lives of purpose and meaning, we live lives of hope. When we base our life decisions on common sense rather than conventional wisdom, we live lives of hope.

Hope is the possibility that something good "could" happen. It is this hope that gives us the strength to challenge the conventional wisdom, to disrupt the status quo, and to advocate revolution. It is this hope that makes us the agents of change, that gives us the ability to communicate our vision to others, and helps us move the world nearer the tipping point of revolutionary change. Even if logic and rationality indicate that our cause is hopeless, our common sense tells us that something good can happen. Our common sense tells us never to give up hope.

Vaclav Havel concluded, "Life is too precious to permit its devaluation by living pointlessly, emptily, without meaning, without love and, finally, without hope."[10] Even if the revolution isn't won in our lifetime, even if is never won at all, life is still is too precious to permit it to be lived as an empty, pointless, meaningless, physical process – of birth, life, death, and decay. Life is too precious to live without hope.

It seems like a long time since I awoke from that restless dream and discovered that my heart had stopped beating for a while. The personal transition had begun years earlier, but the day I came back to life will always remain in my mind as a major turning point in my life. It was the day when I realized I had something important left to do with my life, and I had better be getting on with it. I knew I was going to live for at least a while longer, but I wasn't going to live forever.

Since that day, I have lived a life of hope. I have been healthier than I had been in years. Until the age of 66, until I wore out the fat pads on my feet, I jogged nearly twenty miles a week. I still walk almost an hour a day, I work around the house and yard, and I usually cut at least part of my winter wood supply. Each year, my brother and I find time to ride our bicycles on the Katy trail, stopping at the wineries along the way. I was checked out in the hospital two years after my operation, and pronounced in exceptional physical condition for my age. I have a slight heart murmur, likely from many years of stress, but my heart seems to be working well with no indications that my arteries are re-clogging. Apparently, I still have a while to live and a few more things to do before I go.

My life since has not always been a life of joy or optimism, and at times, the odds of success still seem slim. I sometimes awaken in the morning and wish that my work on earth was about done. But I have never lost the ability to continue working and pursuing my new found purpose in life. I have remained convinced that what I am doing makes sense, regardless of how it turns out. I have learned to live with hope.

My common sense, not logic or reason, tells me that what I am doing makes a difference in the world. My common sense also tells me that everyone makes a difference, regardless of what his or her particular purpose in life may be. My purpose in writing this book is that of a sower. I have attempted to sow the seeds of a revolution of ideals – of new principles, ethics, and measures of success. I have identified the enemy, I have painted a positive vision for the future, and I have pointed out the necessity and inevitability of our ultimate victory.

This is not my revolution alone to fight and win. Odds are I won't live long enough to celebrate the victory. But this revolution is mine to help begin. My job is to sow the seeds of revolution, but even in this, I am far from alone. The seeds of this revolution are being sown by millions of people from all walks of life and from all around the world. All I have done here is gather some bits and pieces of wisdom from these revolutionaries, by whatever means they have come my way. I have attempted to prepare a seedbed and plant those seeds in the fertile ground of open minds. That's my purpose in life. I am a sower. I sow the seeds of revolution as I make the case for a return to common sense.

THE SOWER
The sower sows seeds of wisdom
Wisdom not his own
Wisdom gleaned from others' fields
Harvests from crops God has grown
The sower's harvest season
Comes when the planting ends
Time now to wait for God to grow
New crops of seed to plant again

But the ground grows hard from too much rain
And cracks from too much sun
At times, there's not much to show
For all the sower's done

A few seeds may have opened
A few frail seedlings here and there
But were it not for weeds and grass
The field would look brown, dead, and bare

Should he crack the crust, attack the weeds,
Or plow and plant again?
Will there ever be a harvest?
Or has his planting been in vain?

But the sower's work is over
'Til growing season's end
Then will be time to gather seeds
But just enough to plant again

For the sower's time of harvest
Comes when the planting's done
That's when he reaps the gift of faith
That God's work has begun

The sower's work is man's work
He sows in faith that crops will grow
The miracles of life and growth
Are beyond man, for God to know

The sower leaves the struggling crop
Moves on to plant again
He sows in faith that God will do
What can't be done by man

(John Ikerd)

Jesus used the sower in several of His parables. In Matthew (13:3-10) he said, "A sower went out to sow. And as he sowed, some seeds fell along the path, and the birds came and devoured them. Other seeds fell on rocky ground, where they had not much soil, but when the sun rose they were scorched; and since they had no roots, they withered away.

Other seeds fell on good soil and brought forth grain, some a hundred-fold, some sixty, some thirty. He who has ears, let him hear."

Some of the seeds that I have sown in writing this book have fallen along the path – devoured in the rewriting and editing processes. Some of the seeds successfully sown in the book are destined to fall on the rocky ground of hardened minds. Some will never ask why, and thus will never feel a need for new ideals. The seeds that fall on hardened minds will fail to take root and will wither away under the heat of day-to-day life. But, others will fall on the fertile soil of common sense – on insightful minds that are not afraid to ask why. The seeds that fall on fertile minds will bring forth a manifold harvest – a harvest of new hope that will grow into an epidemic of change, a revolution of new ideals – of happiness, sustainability, and quality of life.

Endnotes

[1] Margaret Meade, *The Quotation Page*, <http://www.quotationspage.com/quote/33522.html> (accessed September, 2006).

[2] Thomas Paine, *Common Sense* (Mineola, NY: Dover Publications, 1776, republished 1997).

[3] Patrick Henry, "Liberty or Death," in *Great American Speeches*, edited by Gregory Suriano (New York: Random House Publishing, 1993).

[4] *Wikiquote*, Susan B. Anthony, "Campaign for Divorce Law Reform," <http://en.wikiquote.org/wiki/Susan_B._Anthony> (accessed September 2006).

[5] Malcolm Gladwell, *The Tipping Point* (New York: Little, Brown, and Co., 2000).

[6] Alexis de Tocqueville, *Democracy in America* (New York: Bantam Books, 2000, original copyright, 1835).

[7] Paul Ray and Sherry Anderson, *The Cultural Creatives: How 50 Million People Are Changing the World* (New York: Three Rivers Press, 2000).

[8] Paul Hawken, *The Ecology of Commerce* (New York: HarperCollins, 1993).

[9] Vaclav Havel, *Disturbing the Peace* (New York: Random House, 1990), 181-182.

[10] Havel, *Disturbing the Peace*, 188.

Index

Aburdene, Patricia, 138, 139, 146
Age of Enlightenment, 11, 79, 80, 81, 90, 133
Age of Insight, 11
Age of Reason, 11, 27, 90, 163, 166, 167
Agricultural Economics, 20, 26, 38, 39, 55, 83, 105, 119, 191
American Agricultural Economics Association, 119
American Revolution, 2, 18, 156, 205, 207
Anderson, Sherry, 209, 214
Anthony, Susan B., 205, 214
Awakening, 1ff, 167

Bill of Rights, 102, 107, 183, 190
Biotechnology, 65, 140, 191
Boundaries: cultural and political, 195, 199, 200; defined, 195-96; ecological, 195; economic, 195-196, 200; families, communities, states, 200; living system, 199; natural, 195, 198, 199; semi-permeable, 200
Bowling Alone, 107, 118
Brane theory, 93
Breimyer, Harold, 44
Bush, George H. W., 16
Bush, George W., 180

CAFOs, 147-149
Campaign finance, 182; information versus persuasion, 183; public financing, 183; reform, 62, 183, 184; right of free speech, 183
Capitalism, 27, 29; defined, 30; fall of, 39ff; destruction of by advertising, 44; replaced by corporatism, 52, 57, 58, 60, 61, 66; restoration of, 118; socialistic restraints, 51-52
Capra, Fritjof, 97, 104
Carson, Rachel, 69, 78
Carter, Jimmy, 73, 77, 78; "crisis of confidence" speech, 73
Cartwright, Edmund, 28

Chaos theory, 81, 95-98, 104, 151
Chaordic organizational paradigm, 99-101, 104
Chrematistics, 120, 121, 128
Christianity, 205
Civil Service, 32
Clinton, Bill, 16, 135
Cobb, John, 120, 132
Collective purchases, 111-112, 114, 115, 118
Common sense, 7; defined, 8; objective reality, 135; philosophy of, 7-8
Common Sense, 2, 14, 154, 156
Communist Manifesto, The, 50, 54
Community supported agrilculture (CSA), 175
Comparative advantage, 197-198
Confinement animal feeding operations. *See* CAFOs.
Constitutional amendment: clean and healthy environment, 117, 187; food, clothing, shelter, health care, 187; elimination of corporate personhood, 186; intergenerational rights, 187; natural persons, 187; protection from economic exploitation, 187
Constitutional convention, 187; Article V - U.S. Constitution, 186
Conventional wisdom, 8, 10; defined, 8-9
Corporate: aristocracy, 149; oligarchy, 207; personhood - 14th amendment, 185; welfare, 185
Corporation: family, 59; not people, 60; publicly held, 59
Corporatism: control over government, 155-56; defined, 59; elections, 62-63; rise of, 55ff; scandals 2002, 72; societal threats, 77-78; society, 61; the process, 57-58; universities, 63-65
Creation story, 165-66; knowledge of good and evil, 165

215

True value, 129-31
Tyranny of majority, 207

U.S. Constitution, 102, 107-110, 128,
    190; 14th Amendment, 185; Bill of
    Rights, 102, 107, 183
U.S. Supreme Court, 60, 62, 155, 183
United Nations (UN), 68, 145, 193, 201
Universal cycle theory, 133
University of Georgia, 55, 105
University of Missouri, 19, 49, 101, 106,
    129

Victory: call to, 147; global, 191; hope
    for, 203; mental, 168-72; personal,
    161; physical, 173-76; societal, 177;
    spiritual, 161-68
Villain: economic, 39; organizational,
    27; societal, 55; the challenge, 67
Visa International, 99, 140
Vision: economic, 119; of change, 133;
    of shared hope, 145; organizational,
    91; personal, 79; societal, 105
Voltaire, 79, 85
Voter apathy, 180

W. K. Kellogg Foundation, 101
Watt, James, 28
Wealth of Nations, 28, 30, 38, 40
Welfare, 75
Whitney, Eli, 28
Wilson & Co., 27
Wilson, Woodrow, 32
Work of Nations, The, 138, 146
World Bank (WB), 193, 194, 201, 202
World Trade Organization (WTO), 193,
    194, 195, 196, 198, 200-202
Worldview, 79, 80, 87, 91, 92, 97, 99,
    137, 141, 142, 193

*Zen and the Art of Motorcycle
    Maintenance*, 135, 146

Printed in the United States
131305LV00003B/91/A